Data handling — 42

🖙 Pie charts — 44

🖙 Scatter diagrams and correlation — 46

🖙 Finding averages — 48

    The mode, median and mean — 48

🖙 Cumulative frequency — 50

Algebra — 52

🖙 Number patterns — 54

🖙 Sequences — 56

🖙 Investigating patterns 1 — 58

🖙 Investigating patterns 2 — 60

🖙 Multiples, factors and prime numbers — 62

🖙 Solving equations with an unknown number on one side — 64

🖙 Solving equations with unknown numbers on both sides — 66

🖙 Solving equations with brackets — 68

🖙 Solving equations with fractions — 70

🖙 Factorising — 72

🖙 Changing the subject of a formula — 74

Formulae with x^2 and negative x terms — 76

🖙 Simultaneous linear equations — 77

Probability — 80

Probability of a single event — 82

🖙 Listing outcomes from events — 83

Mutually exclusive events — 84

The 'OR' rule — 85

Independent events and using the 'AND' rule — 86

🖙 Using tree diagrams — 87

Exam-style questions — 88

Answers to practice questions — 90

About BITESIZEmaths

BITESIZEmaths is a revision guide that has been put together to help you with your GCSE exams. You can watch the TV programmes, work your way through this book and even dial up the Internet on-line service. It's called BITESIZEmaths because that's a good way to revise – in bite-sized chunks, not all in one go.

About this book

It's not possible for us to cover everything that you will need to study in your maths course in the book and TV programmes. Instead, we have concentrated on covering topics that many students find difficult. The book is divided into eight sections, which match the TV programmes. The sections of the book are:

- Using number

- Percentages and fractions

- Measuring

- Shape and space

- Using graphs

- Data handling

- Algebra

- Probability.

Each section has examples that show you how to solve problems, and there are practice questions throughout the book. Work through the sections and make sure you understand the examples before you try the practice questions.

KEY TO SYMBOLS
⑦ A question to think about

◎ An activity to do

ⓘ Work for Intermediate level and above

(TV) A link to the video

⑨ A link to the Internet

Each section also has a Factzone. Here, you will find reminders of maths that you really should know well for the exam. If there are topics in the Factzone that you are unsure about, now is the time to go over your class work again and ask your teacher for advice. Don't be embarrassed about asking for help – most teachers are delighted when they see students taking responsibility for their own learning and their own future.

You will find lots of hints and tips (REMEMBER!) throughout the book but do make your own notes in the margin as well. Write the time code from the video in the relevant section of the book. This will enable you to find that part of the programme more quickly in your next study session.

510

T19300

Graham Lawlor

Published by BBC Educational Publishing,
BBC White City, 201 Wood Lane, London W12 7TS
First published 1998, Reprinted 1999
© Graham Lawlor/BBC Education 1998

ISBN: 0 563 46119 5

Designed by Steve Hollingshead
Printed in Great Britain by Bell & Bain, Glasgow

Contents

BITESIZEmaths

Introduction   4

Using number   8

Using number to solve problems   10
　📺　Money and time   10
　　Reading two-way tables   11
Standard form   12

Percentages and fractions   14

Working with fractions 1   16
　　Equivalent and improper fractions   16
　　Mixed numbers   17
Working with fractions 2   18
　　Adding fractions   18
　　Subtracting, multiplying and dividing fractions   19

Measuring   20

📺 Circle facts   21

📺 Measuring to the nearest unit   22

📺 Calculating the area of shapes   24

📺 Finding the volume   26

Shape and space   28

Using angle facts   30

Using Pythagoras' rule   32

Using Pythagoras' rule to solve problems   34

📺 Sines, cosines and tangents   36

Using graphs   38

　　Conversion graphs   38

📺 Straight line graphs   40

　　Simultaneous linear equations   41

About GCSE

Your school or college will enter you for your maths exam. There will probably be two written papers, and you might also have to complete coursework. Some students have an aural exam, too. It depends which exam board sets the exams that you are going to be taking, but your teacher will tell you what sort of exams you need to prepare for. In an aural exam, a teacher reads questions aloud and you have to work out the answers. You are not usually allowed to use scrap paper and you are not allowed to use a calculator.

Maths exams are different from other subjects because you can enter at three levels: Foundation, Intermediate or Higher. Your teacher will advise you of the level that is best for you – the majority of students enter at the Foundation or Intermediate level. At times in the book, you will see the Intermediate symbol $\boldsymbol{i}$ (usually next to the heading at the top of the page). This means that only students working for the Intermediate and Higher level exams need to cover this work.

Planning your revision

The secret to success is planning. When you have work to learn, you need to go through it several times – it's a bit like learning to drive a car. You wouldn't expect to get into a car and drive it perfectly first time – you need to practise. Learning work for school is the same and that is certainly true in maths.

Plan your time well. Your teacher can advise you on ways to plan your revision. You need to look back over your work regularly. A review schedule like this might be helpful.

Review schedule

Go back to a topic after:

- 10 minutes
- 1 day
- 3 days
- 1 week
- 3 weeks
- 1 month
- 3 months.

Plan your revision three months before the exam. Use your diary and make sure you review your work according to this time schedule. It really does work and will make a huge difference to your success in maths.

In maths, there are two key things that you need to learn. First, you need to learn the material – what 'area' is, for example, or how to find the

circumference of a circle. Don't just read back through your work. Tape the programmes off the TV and watch them as often as you like. Make sketches, summaries of your notes, balloon bubbles and spider diagrams. Use mind mapping and any other technique you can think of to help you remember. Diagrams and pictures are easier for the brain to remember than words, so use lots of pictures, diagrams and codes. This example (right) is a diagram for revising circles.

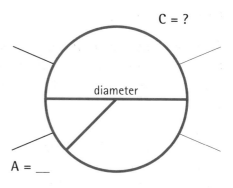

The next thing you need to do is make sure that you understand the techniques you have been taught. Ask your teacher if you can write extra notes in your exercise book. When you have worked through some questions in class, make notes to remind you what to do.

Practise answering questions as part of your revision. Nearer to the exam, get hold of some old exam papers – your teacher should be able to help you with this. You'll find that the same type of question comes up year after year. Obviously, you will not be asked *exactly* the same questions, but looking at old exam papers will give you an idea of the kinds of question to expect. You'll see that some questions are split into several parts, and the marks for each part are shown on the paper.

Getting ready for the exam

Make sure you know the date and time of the exam and where it is going to be held. Every year, some students arrive at the wrong exam room, or turn up in the afternoon to find the exam took place in the morning. Make sure this doesn't happen to you!

Find out how long the exam is going to last, and how many questions you have to answer in that time. Get an idea of how long you can spend on each question.

Make sure you know how to use your calculator. If it is a scientific calculator, the display must be in degrees. This will be shown as D or DEG on the display. This is very important because if it is in a different mode, such as radians and gradians, it will give you the wrong answers. Find out how to get it back into degrees if you accidentally put it into the wrong mode. Non-scientific calculators work on a different operating system. If you use a programmable calculator, you will probably have to clear the memory banks for the exam. If you need help with your calculator, ask your teacher.

When you use the calculator, try to have a rough idea of the answer you would expect. For instance, 3.142 x 30 would give you an answer of about 90, because 3 x 30 is 90. If you got an answer of around 900, you would know that you had made a mistake somewhere.

On the day

Arrive in good time (but not *too* early). Make sure you have a pen (it's a good idea to have two), a ruler and your calculator. When you get into the exam room, stay calm. If you feel a bit panicky, take some deep breaths to help you relax.

Read all the instructions on the front of the paper and fill in your name and exam number. Read the questions carefully. If you are unsure about what you have to do, you can ask the person supervising the exam (who is called the invigilator) for advice, although obviously he or she will not be able to help you answer the questions.

Usually, for the first paper you have to write actually on the paper. For the second paper, you will get an answer book to write in. On the first paper, look at the amount of space that you have been given to write your answer. If there is only a small space, the examiners are only expecting to see an answer, but if there is a bigger space, they will expect you to show your working out. You get marks for showing your working out. Write it clearly so that the examiners can see what you have done. If you show your working out correctly but then make a mistake in the final answer, you can get still get some marks.

Check your time during the exam and don't spend too long on one question. If a question is taking too long, leave it and move on to the next one. You should be able to make up time on questions that you find easier, so at the end of the exam there might be time to go back to the question you missed out.

Remember – an exam is like a performance on stage. You need to rehearse and practise, so that on the day, your performance will be brilliant.

Good luck!

Acknowledgements

The author would like to thank his wife, Judith, for all her support. Thanks also to Jane Furlong of Soar Valley College, Leicester and David Hodgson, member of examining board subject committees and assistant examiner, for their help on this book.

THE ON-LINE SERVICE
You can find extra support, tips and answers to your exam queries on the BITESIZE Internet site. The address is http://www.bbc.co.uk/education/revision

Using number

This section is about:

- using the four basic rules of number

- calculating money and time

- reading two-way tables

- using standard form to write numbers

The four basic rules of number are addition, subtraction, multiplication and division. When you are working on problems, you probably use the words 'add', 'minus' or 'take away', 'times' and 'divide'.

In the exam you will have to work out which rule you need to use. You also need to understand that addition and subtraction are opposites - they undo each other. Multiplication and division are opposites too - they also undo each other.

Estimating is an important skill. In real life, you often need to make estimates (guesses based on the information you have) and then test them. In the exam you may need to make estimates, too.

A common type of question is a drawing of two objects, with measurements given for one but not the other. You are asked to estimate the height of the second object. You need to compare the two and ask yourself questions, such as 'Is one half of the other? A quarter of the other?', and so on. Use the information that you have to make a judgement. The examiner will have a range of acceptable answers, because for estimating there isn't usually an exact answer.

Practise estimating distances. Look at the room you are in now. How many metres long is it? Measure the room and see how close your estimate was. One way of estimating length is to compare the length you want to work out with a length that you already know. For instance, if you know a man who is about 2 metres tall, imagine him lying down in the room. How many times would you be able to fit him into the length of the room?

There is a famous song called 'Twenty-four hours from Tulsa': this shows another way of measuring distance – in terms of the time it takes to get to a place. People often estimate distance in terms of time. For instance, a place might be a 'fifteen-minute bus ride' away, or 'twenty minutes in the car'.

You need to practise estimating time. A good way of estimating seconds is to say to yourself 'one elephant, two elephants, three elephants' and so on. It takes most people about one second to say 'one elephant'.

It is important to practise using number in solving problems. Think the question through and make sure that you understand it. If you are not sure what a question is asking, discuss it with your teacher.

FactZONE

Money

In most currencies, the main unit is divided into 100 smaller units, e.g.

UK: £1 = 100 pence

France: 1 French franc = 100 centimes

USA: $1 = 100 cents

Interest

This is money paid on savings and loans. Interest on savings is paid *by* the bank, building society or Post Office to savers who bank there. Interest on loans is paid *to* banks, etc, by people who have borrowed money from them.

Kilo means 1000

1 kilometre (km) = 1000 metres (m)

1 kilogram (kg) = 1000 grams (g)

so, 2.4 kg = 2400 g

Imperial measurements

Imperial measurements are the old British measurements:

12 inches = 1 foot (roughly 30 cm)

3 feet = 1 yard

1760 yards = 1 mile

16 ounces (oz) = 1 pound (lb)

14 pounds = 1 stone

Weight equivalents

1 kg is about 2 lbs

Distance equivalents

1 mile = 1.609 km (so 5 miles is roughly 8 km)

Time

60 seconds = 1 minute

60 minutes = 1 hour

24 hours = 1 day

7 days = 1 week

52 weeks = 1 year

365 days = 1 year (366 days every 4th year, known as a leap year)

100 years = 1 century

24-hour clock

In the 24-hour clock, 1 pm is shown as 13:00, 2 pm is 14:00, etc.

So 15:30 is 3:30 pm, or half-past three in the afternoon.

17:45 is 5:45 pm or quarter to six in the evening.

Time zones

The world is divided into different time zones. In large countries, such as the USA, there are several time zones within one country.

The time in Greenwich in London is called Greenwich Mean Time (GMT). In some countries, the time is several hours ahead of the time in London. In other countries it is several hours behind. For example, Cairo in Egypt is 2 hours ahead of London, so when it is midday in London it is 2 pm in Cairo. New York, USA, is 5 hours behind London, so when it is midday in London, it is 7 am in New York.

If you are given the time in one country and asked to work out the time in another country, you need to add or subtract the time difference from the time given in the question. For example, Mumbai in India is $5\frac{1}{2}$ hours ahead of London, so if it is 10.00 am in London, it is 3.30 pm in Mumbai $(10 + 5\frac{1}{2}$ hours).

Using number to solve problems

📺 Money

A meal in a restaurant costs £9.70 per person. How many people were in a group who paid £48.50 in total?

(?) *Look carefully at the question. What are you being asked to do?*

You are being asked to find out how many people were in the group.

Step one: Write down what you know
Total amount spent = £48.50
Cost per person   = £ 9.70

Step two: Decide on the maths you need to use
In this case you need to divide:
£48.50 ÷ £9.70 = 5

Step three: Make sense of the answer
The number of people in the group is 5.

(?) *How could you check your answer?*

One way is to do the opposite of dividing – you can multiply:
Number of people (5) x cost per meal (£9.70) = total cost (£48.50)
Check that this works on your calculator.

📺 Time

My grandfather died three years ago at the age of 93 and left me his clock in his will. The clock was made 15 years before he was born.

◎ *When was my grandfather born? When was the clock made?*

Step one: Write down what you know
The year now is 1998. Grandfather died three years ago, so he must have died in 1995. He was 93 when he died.
The clock was made 15 years before he was born.

Step two: Decide on the maths you need to use
In this case you need to subtract:
1995 - 93 = 1902 (year he died minus age)
1902 - 15 = 1887 (year he was born minus age of clock when he was born)

Step three: Make sense of the answer
Grandfather must have been born in 1902.
The clock must have been made in 1887.

Reading two-way tables

You have to be able to read two-way tables and use the information contained in the tables to solve problems.

	Bristol	Cambridge	Cardiff	Edinburgh	Liverpool	York
Bristol		148	45	379	183	227
Cambridge	148		179	334	186	154
Cardiff	45	179		394	198	243
Edinburgh	379	334	394		221	199
Liverpool	183	186	198	221		100
York	227	154	243	199	100	

This table gives distances in miles between some cities in the UK. The same cities are listed across the top and down the side. To find the distance between two places, find your starting point along the top row. Find the place you want to go to on the left-hand side. Read down the column from the top and along the row from the side, until you find the box where they meet. Some of the boxes are blanked out, because there is no distance between York and York, for example.

A group of tourists from the USA are planning a trip around the UK. Starting at Edinburgh, they want to travel no more than 200 miles, then take a break for some sightseeing and stay overnight in this city. On the second day, they want to visit Cambridge and then stay in Bristol.

◎ *Where would you recommend that they stop for the night?*

Starting at Edinburgh, you have to recommend that they go to York, 199 miles away. This is the only city less than 200 miles away.

◎ *How far will they travel during the second day ?*

York to Cambridge	=	154 miles
Cambridge to Bristol	=	148 miles
Total	=	302 miles

There are some practice questions for this section on page 13.

Standard form

Standard form (or standard index form) is a way of writing very large or very small numbers. Most calculators use standard form and present it like this:

4.5 E 07

or

4.5^{07}

This means $4.5 \times 10^7 = 45\ 000\ 000$.

Most textbooks will say something like, 'When a number is written in standard form, it is in the general form $a \times 10^n$ where $1 \leq a < 10$ and n is an integer (whole number).'

This means that a cannot be less than 1 or be 10 or more.

Example 1

◎ *Write 3000 in standard form.*

$3000 = 3 \times 1000 = 3 \times 10^3$

In this example we have changed the 1000 into a power of 10 and so to make 3000 we have to multiply the 10^3 by 3, giving us the answer of 3×10^3.

Example 2

◎ *Write 240 in standard form.*

$240 = 2.4 \times 100 = 2.4 \times 10^2$

Example 3

ℹ *Write 0.0007 in standard form.*

In this example, 0.0007 is written as $7 \times \frac{1}{10000}$. 10 000 would be written as 10^4 in standard form, but we have $\frac{1}{10000}$, which is $\frac{1}{10^4} = 10^{-4}$

so, $0.0007 = 7 \times \frac{1}{10000}$

$= 7 \times \frac{1}{10^4}$

$= 7 \times 10^{-4}$

Practice questions – using number

1) There are 472 students in a college. There are 152 female students. How many male students attend the college?

2) How much do I need to pay for 12 litres of paint, if one litre costs £3.40?

3) A bus leaves Betws-yn-Rhos at 15:30 hours and is due to arrive in Llandudno Junction 45 minutes later. What time will the bus arrive at Llandudno Junction?

4) I spend £14.20 in a shop and give the sales assistant a £20 note. How much change will I get?

5) Which four coins make a total of £1.27?

6) A clock stopped at 07:15 on Thursday and was restarted at 13:20 on Friday. How long did the clock stop for?

7) In 1975 a time capsule was buried for a period of 50 years. In what year is it due to be opened?

8) BBC 1 Schedule

6.00 pm	The Six O'clock News
6.30 pm	Regional News
7.00 pm	Weekend Watchdog with Anne Robinson
7.30 pm	Top of the Pops
8.00 pm	Vets in Practice
8.30 pm	Children's Hospital
9.00 pm	News
9.30 pm	Dangerfield
10.20 pm	Parkinson

a) Martin watches TV from 6.15 pm until the end of *Dangerfield*. How long did he spend watching TV?

b) How long is the *Dangerfield* programme?

9) Find two numbers that add up to 13 but multiply to equal 36.

10) Find two numbers that multiply to 63 but subtract to equal 2.

Practice questions – standard form

Write the following numbers in standard form.

1) 250   **2)** 565

3) 56 000   **4)** 50

5) 3600   **6)** 270

7) 58 000   **8)** 29

9) 850 000   **10)** 3650

11) 0.0008   **12)** 0.00003

13) 0.00055   **14)** 0.367

15) 0.000077   **16)** 0.09

17) When the number 3.95×10^{18} is written out in full, how many 0s will follow the 5?

18) If $x = 2 \times 10^{12}$ and $y = 3 \times 10^5$ find a) xy b) $\frac{y}{x}$.

19) Work out $(3.1 \times 10^{-4}) \times (4.3 \times 10^{12})$.

20) Work out $(3.7 \times 10^{-6})^2$.

21) Given that $v = u + at$, find t in standard form, when $v = 14\ 000$, $u = 0$, $a = 2$.

22) Given that $M = \sqrt{\frac{n}{p}}$ find M in standard form when $n = 3.6 \times 10^2$ and $p = 1.6 \times 10^4$.

23) A virus is 0.000 000 000 000 000 000 457 cm in length. Write this in standard form.

24) The speed of light is 3×10^5 km/s. How many 0s will this number have after the 3?

25) The population is about 60 million. Write this number in standard form.

26) Work out $(1.4 \times 10^6)^2$.

Percentages and fractions

This section is about:

- what a percentage is

- using percentages to calculate things such as VAT

- equivalent fractions

- improper fractions and mixed numbers

- adding, subtracting, multiplying and dividing fractions

You need to be familiar with both percentages and fractions because they occur in other areas of mathematics. For example, you might need to use fractions to answer questions about pie charts.

 In this question, you are asked to work out what fraction of the day is spent working. Here, you would need to know that there are 360° in a circle and be able to work out that 120° is $\frac{1}{3}$ of the circle. An alternative question could be 'What percentage of the day was spent working?' Again, it is important to understand how percentages work and how to apply them in other areas of mathematics. There is more information about percentages in the TV programmes.

You must make sure that you understand fractions thoroughly, but as a back-up, remember that many scientific calculators have a fraction button. Some students don't realise that the a b/c button on their calculator is in fact a fraction button. You may find that the a b/c button on your calculator is a second function or inverse. The way to tell this is to look at where the a b/c is written. If a b/c is

written on a button, you simply press the button. If it is written above a button, you need to press either Second Function (usually written as 2nd F) or Inverse (usually written as INV).

Try entering 1a b/c 2 on your calculator. You should get a display that looks something like 1⌐2, this is $\frac{1}{2}$. If you enter $2\frac{1}{2}$, you need to key it as 2a b/c 1a /bc 2. This will show you a display of something like 2⌐1⌐2, which reads as $2\frac{1}{2}$.

Make sure that you can answer questions that ask you to work out a discount. For example, if apples are sold in packs of 15 and there is a special offer giving you 20% extra apples free, how many apples are in the special offer packs? Discount questions like this are very common, especially at Foundation and Intermediate level.

You might also be asked to work out percentage increases. For example, if a carpet was £4.20 per square metre and the price goes up by 15%, what is the price per square metre now? The £4.20 was 100%, so adding 15% means that you now need to find 115% of the cost. If you are not sure how to do this, ask your teacher for advice. The exam board that uses aural assessment sets questions like this in the aural exam for Higher-level students.

Families of fractions

Family of halves $\frac{1}{2} = \frac{2}{4} = \frac{3}{6} = \frac{4}{8} = \frac{5}{10} = \frac{6}{12} = \frac{7}{14}$ and so on

Family of thirds $\frac{1}{3} = \frac{2}{6} = \frac{3}{9} = \frac{4}{12} = \frac{5}{15} = \frac{6}{18} = \frac{7}{21}$ and so on

Percentages, fractions and decimals

Per cent means out of 100.

$20\% = \frac{20}{100} = \frac{1}{5} = 0.2$ $\qquad$ $25\% = \frac{25}{100} = \frac{1}{4} = 0.25$

$33\frac{1}{3}\% = \frac{33\frac{1}{3}}{100} = \frac{1}{3} = 0.33\dot{3}$ $\qquad$ $50\% = \frac{50}{100} = \frac{1}{2} = 0.5$

$66\frac{2}{3}\% = \frac{66\frac{2}{3}}{100} = \frac{2}{3} = 0.66\dot{6}$ $\qquad$ $75\% = \frac{75}{100} = \frac{3}{4} = 0.75$

• over the final number means 'recurring'

To change from a percentage to a decimal:

Divide the percentage quantity by 100,
e.g. to show 50% as a decimal:

$\frac{50}{100} = 0.5$

To change from a fraction to a percentage:

Multiply the fraction by 100,
e.g. to change $\frac{4}{5}$ to a percentage:

$\frac{4}{5} \times 100 = 80\%$

To find a percentage of a quantity:

Divide the percentage by 100 and then multiply
by the quantity, e.g. to find 15% of 400

$15\% = \frac{15}{100}$

Calculator sequence is 15 ÷ 100 x 400 = 60

Working out VAT

VAT stands for Value Added Tax. In 1997 this
was 17.5% on most goods and services. That
means 17.5% extra is added to the price of
many purchases.

To find 17.5% of £40 without a calculator:

10% of £40 $= \frac{10}{100} \times 40 = £4$

5% of £40 $\qquad$ = £2 (5% is half of 10%, so
answer is half of £4)

2.5% of £40 $\qquad$ = £1 (half of £2)

Total $\qquad$ = £7

Proportion

Measurements that are in direct proportion
increase by the same percentage. This means
that as one increases, so does the other. A
good example of direct proportion is height
and weight. Taller people usually weigh more
than shorter people, because their weight
increases in proportion to their height.

Inverse proportion occurs when one
measurement increases, say, by doubling, but
the other measure decreases by half. Another
example is where a measure increases by three
times but the other measure decreases to a
third of its original quantity.

Working with fractions 1

The top part of a fraction is called the numerator and the bottom part of the fraction is called the denominator.

$\frac{1}{3}$ 1 is the numerator and 3 is the denominator.

$\frac{3}{4}$ 3 is the numerator and 4 is the denominator.

Equivalent fractions

REMEMBER
If you get stuck with fractions, try drawing a diagram.

The diagram shows that $\frac{1}{2}$ is also the same as $\frac{2}{4}$.
It covers the same amount of space in the diagram.

Families of fractions

This is a family of fractions for $\frac{1}{2}$. In other words, each of these fractions is equal and they all equal $\frac{1}{2}$.

$\frac{1}{2} = \frac{2}{4} = \frac{3}{6} = \frac{4}{8} = \frac{5}{10} = \frac{6}{12} = \frac{7}{14}$ and so on

The family is made up of a series, $\frac{1}{2}$, $\frac{2}{4}$. . . The second numerator is generated by 1 x 2, so the denominator must be found the same way, 2 x 2.

Another way to write it is:

$\frac{1 \times 2}{2 \times 2} = \frac{2}{4}$, then $\frac{1 \times 3}{2 \times 3} = \frac{3}{6}$ and then $\frac{1 \times 4}{2 \times 4}$ and so on

◎   *Can you work out the family of thirds?*

You can work out the family of thirds (see Factzone) like this: $\frac{1 \times 2}{3 \times 2} = \frac{2}{6}$ and so on.

Remember, if you multiply the numerator and the denominator by the same number, the value of the fractions is not altered.

Improper fractions and mixed numbers

Improper fractions are 'top-heavy fractions', that is, they are fractions where the numerator is a larger number than the denominator, e.g. $\frac{5}{3}$, $\frac{3}{2}$, $\frac{7}{4}$. Mixed numbers are whole numbers and fractions, such as $1\frac{1}{2}$, $4\frac{2}{3}$.

Changing improper fractions to mixed numbers

◎ *What is $\frac{5}{3}$ as a mixed number?*

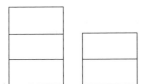

In the diagram (left) you can see $\frac{5}{3}$. In the block on the far left, $\frac{3}{3}$ make up a whole block. There are $\frac{2}{3}$ in the second block, so the diagram shows $1\frac{2}{3}$. In other words, $\frac{5}{3} = 1\frac{2}{3}$.

◎ *What is $\frac{7}{4}$ as a mixed number?*

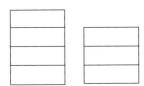

Here, $\frac{4}{4}$ make up the whole block and there are another $\frac{3}{4}$ in the second block, so $\frac{7}{4} = 1\frac{3}{4}$.

Changing mixed numbers to improper fractions

◎ *Change $1\frac{3}{5}$ to an improper fraction.*

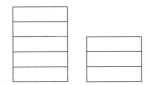

Since the fraction part of the mixed number is in fifths, we need to convert it all to fifths.

You can see from the diagram that the whole block is split into $\frac{5}{5}$ and there are $\frac{3}{5}$ left over. So the answer is $\frac{8}{5}$.

Another way to work this out is $\frac{5 \times 1 + 3}{5} = \frac{8}{5}$.

◎ *Change $3\frac{3}{4}$ to an improper fraction.*

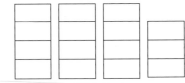

You can see that $3\frac{3}{4}$ is $\frac{15}{4}$

or $\frac{4 \times 3 + 3}{4} = \frac{15}{4}$.

Practice questions

Work out the equivalent fractions.

1) $\frac{1}{2} = \frac{3}{?}$ 2) $\frac{2}{3} = \frac{?}{9}$ 3) $\frac{4}{?} = \frac{20}{25}$ 4) $\frac{1}{4} = \frac{?}{12}$

5) $\frac{3}{?} = \frac{15}{20}$ 6) $\frac{2}{?} = \frac{10}{45}$ 7) $\frac{5}{7} = \frac{15}{?}$ 8) $\frac{7}{8} = \frac{14}{?}$

9) $\frac{5}{?} = \frac{20}{24}$ 10) $\frac{9}{12} = \frac{?}{4}$

Change these mixed numbers to improper fractions.

11) $3\frac{1}{3}$ 12) $5\frac{1}{4}$ 13) $6\frac{3}{7}$ 14) $2\frac{1}{9}$ 15) $9\frac{1}{8}$

16) $3\frac{2}{5}$ 17) $8\frac{2}{11}$ 18) $4\frac{1}{4}$ 19) $2\frac{1}{2}$ 20) $11\frac{3}{10}$

Change these improper fractions to mixed numbers.

21) $\frac{7}{3}$ 22) $\frac{3}{2}$ 23) $\frac{5}{4}$ 24) $\frac{9}{5}$ 25) $\frac{8}{3}$

26) $\frac{12}{8}$ 27) $\frac{11}{5}$ 28) $\frac{9}{6}$ 29) $\frac{15}{12}$ 30) $\frac{21}{8}$

Adding fractions

There are many methods for adding fractions. This is one of the easier methods.

◎ *Add these fractions:* $\frac{1}{2} + \frac{1}{3}$

These fractions have different denominators. You need to change their denominators and make them both the same, before you can add them. This method involves using the families of fractions for $\frac{1}{2}$ and $\frac{1}{3}$.

Step one: Find the lowest denominator that is in both lists

Family of halves $= \frac{1}{2} = \frac{2}{4} = \frac{3}{6} = \frac{4}{8} = \frac{5}{10} = \frac{6}{12} = \frac{7}{14}$ ✓

Family of thirds $= \frac{1}{3} = \frac{2}{6} = \frac{3}{9} = \frac{4}{12} = \frac{5}{15} = \frac{6}{18} = \frac{7}{21}$ ✓

The lowest denominator that is in both lists is 6. In other words, it is the lowest common denominator.

Step two: Rewrite the fractions

$\frac{1}{2} = \frac{3}{6}$ and $\frac{1}{3} = \frac{2}{6}$ so $\frac{1}{2} + \frac{1}{3}$ can be rewritten as $\frac{3}{6} + \frac{2}{6}$.

Step three: Add the numerators

> **REMEMBER**
> Only add or subtract the numerators.

$\frac{3}{6} + \frac{2}{6} = \frac{5}{6}$

That means that $\frac{1}{2} + \frac{1}{3} = \frac{5}{6}$.

◎ *Add these fractions:* $3\frac{3}{4} + 2\frac{1}{3}$

Step one: Split the mixed numbers and add the whole numbers

$3\frac{3}{4} + 2\frac{1}{3} = 3 + 2 + \frac{3}{4} + \frac{1}{3}$

$= 5 + \frac{3}{4} + \frac{1}{3}$

Step two: Find the lowest common denominator for the family of three-quarters and the family of thirds and rewrite the fractions

Family of three-quarters $= \frac{3}{4} = \frac{6}{8} = \frac{9}{12} = \frac{12}{16} = \frac{15}{20}$ ✓

Family of thirds $= \frac{1}{3} = \frac{2}{6} = \frac{3}{9} = \frac{4}{12} = \frac{5}{15} = \frac{6}{18} = \frac{7}{21}$ ✓

The lowest common denominator is twelfths, so $\frac{3}{4} + \frac{1}{3} = \frac{9}{12} + \frac{4}{12}$.

Step three: Add the fractions

> **REMEMBER**
> Where you have to add mixed numbers, add the whole numbers and then add the fractions.

$\frac{9}{12} + \frac{4}{12} = \frac{13}{12} = 1\frac{1}{12}$

Step four: Add this figure to the 5 you worked out in step one

$5 + 1\frac{1}{12} = 6\frac{1}{12}$

so that means $3\frac{3}{4} + 2\frac{1}{3} = 6\frac{1}{12}$.

Subtracting fractions

You can also subtract fractions by using these methods.

◎ *Work out the answer to: $\frac{1}{2} - \frac{1}{3}$*

❓ *What do you need to do first?*

Step one: Find the lowest common denominator in the family of halves and the family of thirds

Family of halves = $\frac{1}{2} = \frac{2}{4} = \frac{3}{6} = \frac{4}{8} = \frac{5}{10} = \frac{6}{12} = \frac{7}{14}$ ✓

Family of thirds = $\frac{1}{3} = \frac{2}{6} = \frac{3}{9} = \frac{4}{12} = \frac{5}{15} = \frac{6}{18} = \frac{7}{21}$ ✓

Step two: Rewrite the fractions and subtract the numerator of the second fraction

So $\frac{1}{2} - \frac{1}{3} = \frac{3}{6} - \frac{2}{6} = \frac{1}{6}$

19

REMEMBER
Don't forget to subtract at the end, instead of adding.

Multiplying fractions

The diagram shows a pie that has been cut in half. Then one of the halves has been cut in half.

This is a simple example of multiplying fractions. In the second cut, we are actually working out the mathematics of $\frac{1}{2} \times \frac{1}{2} = \frac{1 \times 1}{2 \times 2} = \frac{1}{4}$. In other words, we are asking 'What is a half of a half?' As you can see from the diagram, it is a quarter. This explains why multiplying a fraction by another fraction leads to a smaller fraction. Notice that mathematicians say 'half of a half', not 'half times half'.

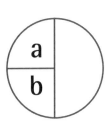

REMEMBER
When you are multiplying, multiply the numerators and the denominators.

Dividing fractions

❓ *What is the effect of dividing a fraction by another fraction?*

In this example we are going to divide $\frac{1}{2}$ by $\frac{1}{4}$. In simple language, this means that we are asking how many quarters will go exactly into a half. In a question like this, it is reasonable to expect the answer to be a whole number.

$\frac{1}{2} \div \frac{1}{4} = \frac{1}{2} \times \frac{4}{1} = \frac{1 \times 4}{2 \times 1} = \frac{4}{2} = 2$

This tells us that there are 2 quarters in a half. Notice that with this technique, you have to turn the second fraction upside down before you multiply.

Look at some other fractions. Make sure you understand what it is that you are calculating.

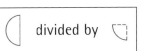
divided by

Practice questions

Work out the following:
1) $\frac{1}{3} + \frac{1}{4}$
2) $\frac{3}{4} \times \frac{1}{5}$
3) $\frac{8}{9} - \frac{1}{3}$
4) $\frac{3}{4} \div \frac{1}{2}$
5) $\frac{12}{9} + \frac{6}{5}$
6) $5\frac{1}{3} + 6\frac{1}{2}$
7) $5\frac{9}{10} - 3\frac{1}{3}$
8) $(\frac{3}{8} + \frac{5}{9}) - \frac{2}{5}$

Measuring

This section is about:

- why we need to measure things

- measuring to the nearest unit

- names of the most common polygons

- calculating areas and volumes

- circles

In the TV programme you saw why we need to measure things. It helps us to describe and to see trends. The programme showed students having their height measured. A question you might ask is, are students taller now than they were thirty years ago? This is not such a crazy question. If people are getting taller, many everyday items will have to be changed. For example, doorways will need to be taller, clothes will have to be made longer, and so on.

In the Factzone there are a number of measurements that you must know. It's worth making the effort to learn them. You might be asked to convert measurements from one system to another, so make sure that you know how to convert from imperial measurements (traditional British units: feet and inches, pounds and ounces, etc.) to metric measurements, and from metric to imperial.

Make sure you know the names of the regular polygons. Intermediate and Higher level

students need to know how to calculate the interior angle of a polygon.

Circles are important at all levels. You must know the names of all parts of a circle. Make sure you use the correct formula for finding a specific area within a circle. If you are given a diameter and asked to find the area, you have to halve the diameter and then put this value into the formula. It is worth spending time to make sure you are familiar with the formulae.

Intermediate level students should make sure they understand the meaning of squaring: x^2 means $x \times x$ and not $2 \times x$.

$A = \pi r^2$ means $A = \pi \times r \times r$.

All students need to know how to round numbers up or down, to a number of decimal places or to a significant figure (such as the nearest unit of 10), because this is regularly required in exams. You also need to practise rounding to a sensible degree of accuracy. If you are unsure about this, ask your teacher.

Measurements

1 km = 1000 m

1 m = 100 cm

1 cm = 10 mm

1 foot = 12 inches

3 feet = 1 yard

1760 yards = 1 mile

1 foot is about 30 cm

1 kg is about 2 pounds weight

1 mile = 1.609 km (so 5 miles is roughly 8 km)

Regular polygons

Quadrilateral: A general word for four-sided shapes, such as squares and rectangles

Pentagon: Five sides

Hexagon: Six sides

Heptagon (also known as septagon): Seven sides

Octagon: Eight sides

Nonagon: Nine sides

Decagon: Ten sides

📺 Circles

Circumference: The distance around the outside of the circle

Diameter: The distance across a circle. It has to be drawn as a straight line from one part of the circumference to another part of the circumference, through the centre point

Radius: Half of the diameter

Chord: The distance across a circle, drawn as a straight line but NOT passing through the centre point

circumference

chord

diameter

radius

Finding the circumference

If you know the diameter, use the formula $C = \pi d$ (π x the diameter).

If you know the radius, use $C = 2\pi r$ (2 x π x the radius).

Finding the area

Use the formula $A = \pi r^2$ (π x the radius x the radius).

Measuring

21

How accurate do you need to be when measuring? The answer is, it depends on what you are measuring.

(?)   *What would be the best unit to use if you were measuring your height?*

If you were measuring your height, it would make sense to measure to the nearest centimetre. But if you were measuring the tolerance of a piece of technical equipment, for example, you would have to measure far more accurately, perhaps to the nearest 10 000th of a millimetre.

Rounding

The rule for rounding or approximating is:

'If it is five or more than five, you round up. If it is less than five, you round down.'

Let's look at some examples to make this clear.

Example 1

This diagram shows the length of the nail.

(◎)   *Look carefully at the diagram. How long is the nail, to the nearest centimetre?*

In other words, which whole number of centimetres is the nail closest to? You can see that it is closer to 2 cm than to 1 cm. So the length of the nail, to the nearest whole centimetre, is 2 cm.

(?)   *How does the rounding rule work in this example?*

The nail is just over 1.7 cm. The last figure (7) is more than 5, so you round up.

Example 2

◎ *To the nearest 10 feet, how long is the van?*

Applying the rounding rule, the last figure (the '3' in 23 feet) is less than 5, so we round down. To the nearest 10 feet, the van is 20 feet long.

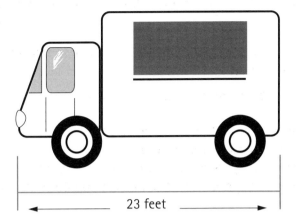

23 feet

Practice questions

1) Look at these objects. Round the lengths of the objects shown to an appropriate unit.

a)

c)

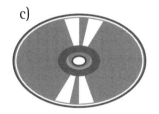

◄— 11.7 cm —►

b)

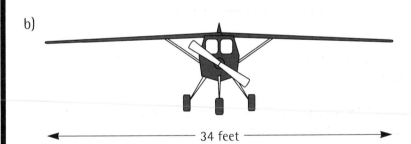

◄———————— 34 feet ————————►

d)

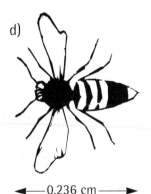

◄— 0.236 cm —►

2) Martin measures the length of a garden fork. The garden fork is 98 cm long.

a) Estimate the length of the garden fork to the nearest foot.

b) Estimate the length, in inches, of 3 forks laid end to end.

3) Estimate how many pounds is 12 kilos of sugar, to the nearest pound weight.

4) Estimate the length of a 24-inch laser printer, to the nearest centimetre.

5) The LCD display on the laser printer in question 4 is $\frac{1}{8}$ of the total length of the printer. Estimate the length of the LCD display, to the nearest centimetre.

24

Area is the measure of flat space on a 2-D shape. 2-D means 2-dimensional, that is, flat shapes, like the diagrams in this book.

One student wrote his notes like this:

For each standard shape, there is a formula that you can use to find its area.

The area of a rectangle = length x width (this is written as lw in algebra).

The area of a square = length x length (this is written as l^2 in algebra).

The area of a triangle is found by halving the length of the base line and multiplying this figure by the height ($\frac{1}{2}$ base x height).

The area of a circle is found by multiplying the Greek letter π by the square of the radius (πr^2).

⊙ *Use the information in this paragraph to make up your own Factzone. Show all these shapes and the formulae for finding their areas. Make a spider diagram to show this information in a more readable form.*

Rectangles and squares

To find the area of a rectangle, use the formula: $A = lw$.

❗ REMEMBER
Area is in square units, so there has to be a ^{2} in the answer.

A = 25 x 100

A = 2500 m²

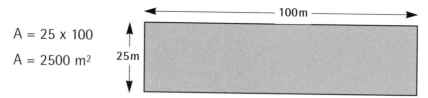

The area of a square is found in a similar way, but you have to use the formula for a square ($A = l \; x \; l$ or l^2).

In the exam, they usually ask you more complex questions, like this one:

The diagram shows the floor plan of an office, marked A on the map, and the storage room that joins on to it, marked B.

a) Find the area of carpet needed to cover the office floor.

b) Find the area of carpet needed for the storage room floor.

c) Hence, find the total area of carpet needed to cover both floors.

❗ REMEMBER
When questions are broken into parts, the first part is usually easier and helps you find the answer to the other parts.

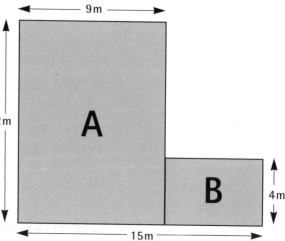

'Hence' is a word that examiners like to use. It means 'so'. In this question, you are being asked to find the area of the first room, then the area of the second room, and so use this information to find the total area.

Step one: Find the area of A

Area of A = 12 x 9

= 108 m²

Step two: Find the area of B

Area of B = 6 x 4 (The 6 comes from 15 - 9 (the width of A + B together minus the width of A))

= 24 m²

Step three: Work out the total area

Total area = 108 + 24 = 132 m²

◎ *Try making up some questions of your own, showing how you work out the area. You can also use this as part of a discussion with an adult. Make sure he or she is clear about what you have done. Not all adults are maths teachers and many may have forgotten how to do the work for themselves.*

Practice questions

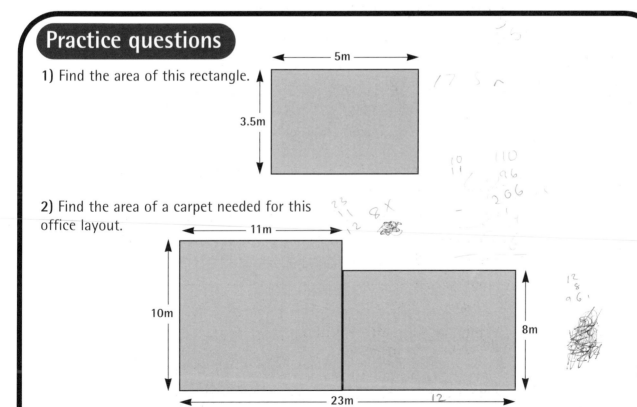

1) Find the area of this rectangle.

5m

3.5m

2) Find the area of a carpet needed for this office layout.

11m

10m

23m

8m

3) Catherine has to paper a bedroom wall that is 3 m high and 10 m long. The wallpaper she wants comes in rolls 2 m wide and 5 m long. How many rolls of paper will she need?

4) Martin wants to paint a fence. The fence is 14 m long and 3 m high. He has to paint both sides of the fence. One tin of paint covers 12m² of fence. How many tins will he need?

ⓘ📺Finding the volume

Volume is a measure of the amount of space inside a 3-D object, like a box or a can. Volume is measured in cubic units, that is, cubic metres (m^3), and cubic centimetres (cm^3). Make sure that you can work out the volumes of standard shapes.

Prisms

A prism is any solid of uniform cross-section. This means that wherever you cut the solid along its length, it is always the same (uniform means the same).

Cylinders

A good example of a prism is a tin of beans. Beans come in cylindrical tins. If you could cut a tin along its length, it would always look the same, no matter where you cut.

Notice that the end of the cylinder (the base) is a circle.

To work out the volume of a cylinder, you can use this formula:

Volume   = area of base x height

         = area of circle (πr^2) x height

❗ REMEMBER
Volume of a cylinder = area of base x height

◎ *Find the volume of a cylinder with a base radius of 4 cm and a height of 10 cm.*

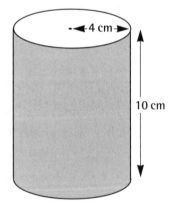

Volume   = area of base x height

         = πr^2 x height

         = π x 4 x 4 x10

         = 502.7 cm^3 (to 1 decimal place)

Triangular prisms

A well-known triangular chocolate bar is often thought to be a prism. The chocolate itself isn't a prism because there are dips and humps along the top of the bar, but the box that it comes in *is* a triangular prism.

In this case the end of the prism is a triangle, so you can use this formula to work out the volume:

Volume = area of triangular end x length

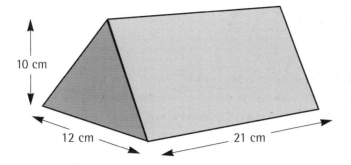

10 cm

12 cm ····· ····· 21 cm

◎ *Find the volume of a triangular prism in which the cross-section triangle base = 12 cm, the triangle height = 10 cm, and the length of the prism = 21 cm.*

Volume   = area cross-section x length

          = area of triangle x length

          = $\frac{1}{2}$ x 12 x 10 x 21

          = 1260 cm^3

◎ *Fill in the missing words:*

    _____ are solids where the cross-section is always _____ _____. This means that it doesn't matter where the solid is cut because the cross-section will always look the same.

    In mathematics, this is called a cross-section that is _____ .

When you are working out volumes, make sure that the units for the height and the units for the length are the same – all in centimetres or all in metres. If they aren't the same, you will have to convert them before you start the calculation.

❗ REMEMBER
Volume of a triangular prism = area of cross-section x length.

Measuring

Practice questions

Remember, with circles or cylinders, to use your calculator π button, or 3.142.

1) Find the volume of a cuboid of dimensions 2.5 m, 4 m and 6 m.

2) A swimming pool is half full of water. If the dimensions of the pool are 9 m by 12 m by 2 m, how many cubic metres of water are in the pool?

3) Find the volume of a cylinder that has a base radius of 10 cm and a height of 12 cm.

4) Julie has a waste bin that is cylindrical in shape. The base has a radius of 0.49 m and the height is 0.82 m. Find the volume of the bin.

5) A kart-racing track has a tunnel shaped like a triangular prism. The dimensions of the tunnel are: 13 m long, 3 m wide and 4 m from its highest point to the ground. Find the volume of the tunnel.

6) A time capsule in the shape of a cylinder is being placed in a wall for twenty-five years. The dimensions of the cylinder are: base radius 12 cm, length 100 cm. The hole in the wall has to have 10% more volume than the cylinder. What will the volume of the hole be?

This section is about:

- using angle facts to solve problems

- using Pythagoras' rule to find missing sides of triangles

- using sine, cosine and tangent ratios to find missing sides and angles in right-angled triangles

Make sure you understand how to solve equations before you start this section.

Angle facts are important in solving problems. Make sure you understand them thoroughly. Practise recognising alternate and corresponding angles. Intermediate and Higher-level students will have to use algebra to solve some triangles. Remember the angles in a triangle add to 180° and use this fact to form an equation. Your equation must make mathematical sense. Remember to treat it the same way as you would treat an equation in algebra. If you're unsure, look at the TV programmes and at the algebra sections in this book.

Intermediate and Higher-level students will have to know how to calculate the lengths of sides and the sizes of angles in triangles. The triangle must have a right angle in it. If there doesn't seem to be a right angle, you will need to draw in a line to create one.

When you are using Pythagoras' rule, remember that squaring means multiplying by itself:

3^2 means 3 x 3 and NOT 3 x 2.

Make sure you know how to use the functions on your calculator. You will find buttons such as $\sqrt{}$, but they may be second functions on your calculator. Check the instruction booklet that comes with the calculator and learn the correct key sequences. This is important in trigonometry, especially when finding 'arc' values (also known

as inverse trig values). Older scientific calculators asked you to put in the value and then find the trig ratio. With most newer calculators you don't need to do it that way: you press the trig function first and then put in the value. For instance, tan 13° in an older calculator is found by keying 13, then the tan button. In modern calculators it is keyed as it appears.

Make sure that your calculator is in the correct mode, DEGREES. You can check this by working out sin 30° ($=\frac{1}{2}$), cos 60° ($=\frac{1}{2}$) or tan 45° (=1).

Pythagoras' rule and trigonometry sometimes cause anxiety and problems for students. Make some extra time to work carefully through this chapter. First of all, you need to understand the ideas. Then, you need to practise using the ideas to answer questions.

Make some charts that you can put up on your bedroom wall and see every day. Don't limit yourself to the examples and practice questions here – make up some of your own and try them out. If you are unsure how Pythagoras' rule works, investigate it. Try some simple cases for yourself, keeping careful note of what you have done. Discuss it with another person. Do the same thing for trig.

You need to understand how transformations work. Make sure you understand reflections, rotations, translations and enlargements. There is information about these in the TV programmes.

Triangles

Solving a triangle: This means finding missing lengths of sides and finding missing angles.

Angles in Triangles

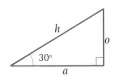

Look at this triangle.

Make sure you understand how the labelling works. The side opposite the right angle is called the hypotenuse, and is marked h on the diagram. It is the longest side of the triangle. The side next to the marked angle is the adjacent side (adjacent means 'next to'). The side that is opposite the marked angle is the opposite side. Labelling sides other than the hypotenuse (which is always the longest side) depends on which angle you are being asked to work with.

The marked angle is 30° and we know one of the other angles is 90° because it is a right-angled triangle. The missing angle must be 60° because angles in a triangle add to 180°.

Looking at the 60° angle, you can see the side that was adjacent for the 30° angle is not the adjacent side for the 60° angle. It is the opposite side.

Labelling sides

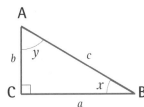

There are two ways of doing this. The first way is to use capital letters at the points, so the hypotenuse of this triangle is AB. The other two sides are labelled AC and BC. The second method is to use lower-case letters and label the side according to the lower-case point opposite the side. So, AB is also lower-case c, AC is lower-case b and BC is lower-case a.

Not drawn to scale: This phrase means that you cannot measure the diagram to find the answer. The diagram is not actually drawn in proportion to the measurements shown.

ⓘ Notation

In trigonometry there are two main types of notation. There is the notation that is usually seen on calculators, such as $\sin^{-1}$, $\cos^{-1}$ and $\tan^{-1}$. These mean 'the angle whose sine is', 'the angle whose cosine is' and 'the angle whose tangent is'. Take care not to confuse this notation with the same notation used in algebra, namely x^{-1}. In algebra this notation means $\frac{1}{x}$ but $\sin^{-1}$ does not mean $\frac{1}{\sin}$.

Arc notation is an alternative notation that more exam boards are starting to use. It means the same thing as inverse trig notation but makes it easier to avoid errors. Your teacher will know which notation you have to use.

Shape and space

Using angle facts

You can use angle facts to find the missing angles in triangles.

Angles on a straight line: Add up to 180°: $x + y = 180°$. You can use this fact to work out missing angles, by subtracting the angle you know from 180°

Angles around a point: Add up to 360°

Add the angles that you have and subtract them from 360°. This will give you the size of the missing angle.

Alternate angles: Commonly known as 'Z' angles. The angles in the crooks of the Z are equal. The Z may be backwards, as in the second diagram, but the angles in the crooks of the Z are still equal.

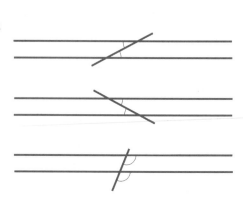

Corresponding angles: Commonly known as 'F' angles. Trace out a capital F on the diagram. It can be facing left or right, or even upside down, but the angles under the 'arms' of the F are equal

◎ *If angle x is 48°, work out the missing angles in the triangle shown.*

In a question like this, use the fact that angles add up to 180°. The marks on the lines show that these two sides are equal in length, so it must be an **isosceles triangle**. From this, we can see that the two missing angles must be equal.

The answer must be: $\frac{180-48}{2} = \frac{132}{2} = 66$.

In other words, each of the two missing angles must equal 66°.

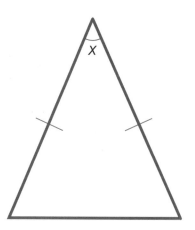

30

◎ *Find the missing angles in this triangle.*

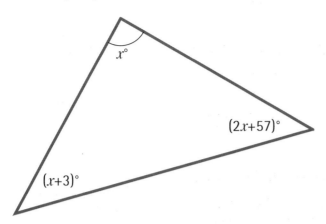

REMEMBER
You can use algebra to solve triangles. Write out the equation first.

31

Again, use what you know about angles in triangles.

In this case we can set up an equation:

$x + (x + 3) + (2x + 57) = 180$

$4x + 60 \qquad\qquad = 180$

$4x + 60 - 60 \qquad = 180 - 60$

$4x \qquad\qquad\quad = 120$

$\frac{4x}{4} \qquad\qquad\quad = \frac{120}{4}$

$x \qquad\qquad\qquad = 30$

Therefore, the missing angles are 30°, 33° $(x + 3)$, and 117° $(2x + 57)$.

Practice questions

1) Find the value of x, in degrees.

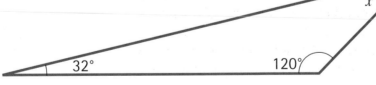

2) The three angles of this triangle are x, $2x$ and $3x$. Find the size of each angle, in degrees. (You will need to set up an equation.)

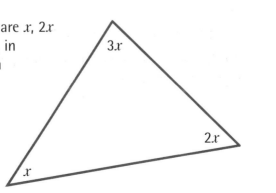

Shape and space

BITESIZEmaths

❶Using Pythagoras' rule

Pythagoras was a Greek who lived around 500 BC. He is credited with having discovered a unique fact about right-angled triangles:

'The square on the hypotenuse of a right-angled triangle equals the sum of the squares on the other two sides.'

This is known as Pythagoras' rule (some older textbooks use the word 'theorem' instead of rule). What exactly does it mean?

Look at the example on the left.

The hypotenuse is the longest side in a right-angled triangle. It is always opposite the right angle. In this triangle, the hypotenuse is clearly 5 cm, since this is the longest side.

'The square on the hypotenuse' means the length of the hypotenuse squared. The square on the hypotenuse equals $5^2 = 25$.

Pythagoras' rule says that this is 'equal to the sum of the squares on the other two sides'. This means the length of each of the other sides is squared and then they are added together.

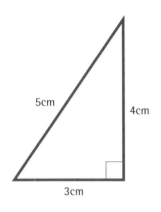

$3^2 = 9$ $\qquad\qquad\qquad\qquad\qquad$ $4^2 = 16$

$9 + 16 = 25$ (the square on the hypotenuse)

You can use this rule to find the missing sides of right-angled triangles.

❗ REMEMBER Finding missing sides and/or missing angles is called 'solving the triangle'.

Finding the length of the hypotenuse

Example 1

Look at this triangle. The missing side is x, so we square it, making it x^2. The other two sides squared are 6^2 and 8^2.

Using Pythagoras' rule, we can form an equation.

$x^2 = 6^2 + 8^2$

$x^2 = 36 + 64$ $\qquad$ Work out the right-hand side of the equation.

$x^2 = 100$

 Is this the answer to the question?

We don't want x^2, we want x. To find this, we need to take the square root of each side of the equation.

$\sqrt{x^2} = \sqrt{100}$

$x = 10$ cm

By using Pythagoras' rule, we have found the length of the missing side.

Example 2

In some triangles, the lengths of the sides are not easy-to-use whole numbers – they involve decimals. The same rule still works. Look at this triangle.

Using Pythagoras' rule:

$x^2 = 7.2^2 + 5.9^2$

$x^2 = 51.84 + 34.81$

$x^2 = 86.65$ Now take the square root of both sides of the equation.

$x = 9.308598176$

$x = 9.31$ cm (to 2 decimal places)

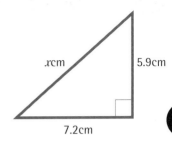

Finding other missing sides

Pythagoras' rule can also be used to find a missing side that is not the hypotenuse. To understand this more easily, let's use a 3, 4, 5 triangle. Here, we have a missing side that is not the hypotenuse.

Using Pythagoras' rule: $5^2 = 3^2 + x^2$

 Try to solve the equation.

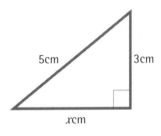

We need to rearrange the equation to get x on its own, with all the other numbers on the other side of the equation. We do this by taking 3^2 away from both sides: $5^2 - 3^2 = 3^2 + x^2 - 3^2.$

On the right side, 3^2 is subtracted, leaving only the x^2. Now the equation reads as:

$5^2 - 3^2 = x^2$ (This is exactly the same as $x^2 = 5^2 - 3^2$)

$25 - 9 = x^2$

$16 = x^2$

$x = 4$ cm

! REMEMBER You have to find the square root of x to find the answer.

Since the triangle is the 3, 4, 5 triangle, this is the result that we would expect.

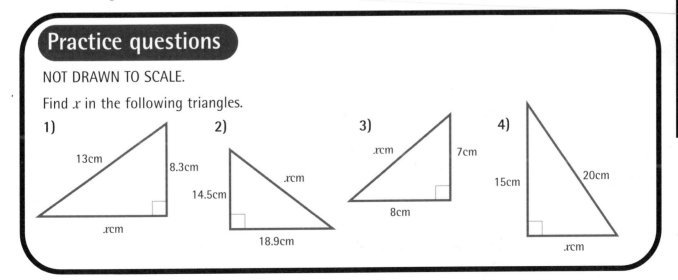

Practice questions

NOT DRAWN TO SCALE.

Find x in the following triangles.

1) 13cm 8.3cm $x$cm

2) 14.5cm $x$cm 18.9cm

3) $x$cm 7cm 8cm

4) 15cm 20cm $x$cm

Using Pythagoras' rule to solve problems

◎ *Find the length of the diagonal of a rectangle that is 8 cm long and 5 cm wide.*

Step 1: Draw a diagram

Drawing a diagram of the shape makes the question easier to understand. You can see from the diagram that this is still a right-angled triangle question, so you simply apply Pythagoras' rule in the same way.

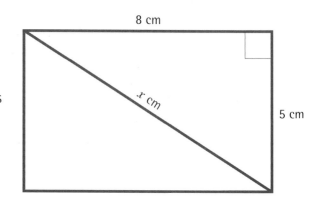

Step 2: Use Pythagoras' rule

Let the diagonal be x cm.

$x^2 = 8^2 + 5^2$

$x^2 = 64 + 25$

$x^2 = 89$ Take the square root of both sides.

$x = 9.433981132$

$x = 9.43$ (to 2 decimal places)

The diagonal is 9.43 cm (to 2 decimal places).

◎ *A square has diagonals 20 cm in length. Find the length of one side of the square.*

Step one: Draw the diagram

Step two: Use Pythagoras' rule

We know that the sides of a square are the same length. Let the sides be x cm.

$20^2 = x^2 + x^2$

$400 = 2x^2$ Divide both sides of the equation by two.

$200 = x^2$ Take the square root of both sides of the equation, putting x on the left-hand side for neatness.

$x = 14.14213562$

$x = 14.14$

The diagonal is 14.14 cm (to 2 decimal places).

❗ REMEMBER Don't finish a number with a full stop. If you write 14.14., for example, it looks as though the number has two decimal points.

Not all triangles have whole numbers to calculate with, so let's consider a different example.

◎ *A rectangle has 1 side of 8.2 cm and a diagonal of 12.4 cm. Find the length of the missing side.*

Step 1: Draw the diagram

Step 2: Use Pythagoras' rule

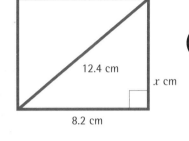

$12.4^2 = x^2 + 8.2^2$

$153.76 = x^2 + 67.24$ Now subtract 67.24 from both sides. This will leave the x^2 term on its own on the right-hand side of the equation.

$153.76 - 67.24 = x^2 + 67.24 - 67.24$

$86.52 = x^2$ Now take the square root of both sides. For neatness put the x on the left-hand side of the equation.

$x = 9.301612763$

$x = 9.30$

The result is 9.30 cm (to 2 decimal places).

Practice questions

Draw diagrams to help you answer these questions.

1) Find the length of the diagonal of a rectangle, with length 10 cm and width 4 cm.

2) A square has diagonals of length 15 cm. Find the length of the sides.

3) A 5 m ladder rests against a wall with its foot 2 m from the wall. How far up the wall does the ladder reach?

4) A plane flies 30 km north and then 40 km west. How far away is it from its starting point?

5) An army platoon travels 42.43 km on a bearing of 045 degrees. Then they travel on a new bearing of 180 degrees for a distance of 30 km. How far away are they from their original starting point?📺

6) Martha and Prakash were arguing about a triangle that had sides of 9 cm, 12 cm and 15 cm. Prakash said the triangle had a right angle. Martha said it did not. Who was correct?

7) A piece of curtain material is a rectangle of length 2.5 m and width 1.6 m. How long is the diagonal?

8) A design on a book cover has a diagonal from corner to corner measuring 30 cm. The book fits exactly onto a book shelf that is 20 cm below the shelf above. How wide is the front cover of the book?

9) A tin of liquid is a cylinder of radius 10 cm and height 25 cm. A brush is dropped into the liquid and goes under the surface of the liquid. The stick lies under the surface as a diagonal. Work out the maximum length that the stick could be.

10) A lean-to shed has a sloping roof 4 m in length, and it is 3 m wide. At its lowest point the roof is 2 m above ground. How high is the highest point of the roof above ground?

Look at this triangle.

$$\text{Ratio} = \frac{\text{opposite}}{\text{hypotenuse}} = \frac{4}{8} = \frac{1}{2}$$

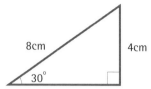

8cm 4cm 30°

◎ Draw a triangle with a hypotenuse of 10 cm, a 30° angle and a side of 5 cm opposite the marked angle. Compare this with the triangle drawn here. What do you notice?

The ratio is the same for both triangles. In fact, any triangle which has a 30° angle in this situation will always have its opposite side half as long as its hypotenuse.

So, there is a ratio of the $\frac{\text{opposite}}{\text{hypotenuse}}$ and this is called the **sine** of the angle.

Sine is shortened to sin when calculating.

There are two other ratios, **cosine** and **tangent**.

Cosine: $\cos = \frac{\text{adjacent}}{\text{hypotenuse}}$ Tangent: $\tan = \frac{\text{opposite}}{\text{adjacent}}$

Finding the length of a side of a triangle

Example 1

REMEMBER

$$\sin = \frac{\text{opp}}{\text{hyp}}$$

Find the length of the side marked x in the triangle.

$\sin 32 = \frac{x}{12}$ Multiply both sides by 12.

$12 \times \sin 32 = x$

$x = 6.359031171$

$x = 6.36$

The length of the opposite side is 6.36 cm (to 2 decimal places).

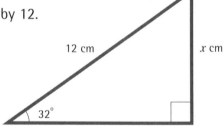

12 cm x cm 32°

Example 2

REMEMBER

$$\cos = \frac{\text{adj}}{\text{hyp}}$$

Find the length of x in the triangle.

❓ *Which rule do you need to use?*

$\cos 45 = \frac{x}{14}$ Multiply both sides by 14.

$14 \times \cos 45 = x$

$x = 9.899494937$

$x = 9.90$

x is 9.90 cm (to 2 decimal places).

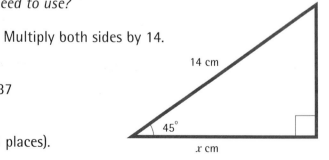

14 cm 45° x cm

Example 3

Find the length of x.

$\tan 60 \quad = \frac{x}{7}$ Multiply both sides by 7.

$7 \times \tan 60 \quad = x$

$x \qquad = 12.12435565$

$x \qquad = 12.12$

The length of x is 12.12 cm (to 2 decimal places).

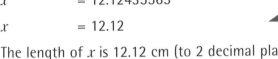
x cm

$60°$

7 cm

Finding a missing angle

On your calculator you will have a function key with either an arc function or $\sin^{-1}$, $\cos^{-1}$ and $\tan^{-1}$. Make sure you understand how to use the calculator. It is essential for the exam. You may have to press a shift or inverse button first.

Example 1

Find the size of angle x.

$\sin \qquad = \frac{\text{opp}}{\text{hyp}}$

$\sin x \qquad = \frac{15}{30}$

$\sin x \qquad = 0.5$ Press arc sin 0.5 (= 30).

$x \qquad = 30°$

30 cm

15 cm

x

Arc sin means 'the angle whose sin is . . .'. You may have used the alternative notation of $\sin^{-1}$.

Example 2

Find the size of angle z.

$\tan \qquad = \frac{\text{opp}}{\text{adj}}$

$\tan z \qquad = \frac{14}{30}$

$\qquad \qquad = 0.46666$ (recurring)

$\qquad \qquad$ Press arc tan 0.466666

$\qquad \qquad = 25.01689348$

$z \qquad = 25.0°$ (to 1 decimal place)

30 cm

z

14 cm

Arc tan means 'the angle whose tangent is'. You may have used the alternative notation of $\tan^{-1}$.

Angles are quite often given to an accuracy of one decimal point.

Shape and Space

Conversion graphs

You can use conversion graphs to convert from one unit to another, for example, from one currency to another, or from metric to imperial measures (feet and inches, pounds and ounces, etc.).

REMEMBER
If you have to draw a conversion graph like this in the exam, make sure you use the exchange rate given on the exam paper.

Example 1

This graph is a conversion graph for British and French money. The exchange rate changes every day. On the day this graph was drawn, the exchange rate was:

£1 = 9.8 French francs

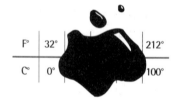

Example 2

Temperature can also be shown on a conversion graph. Fahrenheit and Celsius are two different scales for measuring temperature.

 Someone has dropped ink on this table of values. Use the table to draw the graph. From your graph estimate:

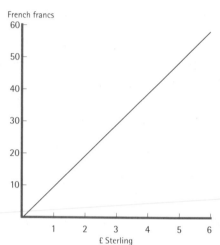

a) What is 60 °F in degrees Celsius?

b) What is 75 °C in degrees Fahrenheit?

c) What values could be in the table, under the ink blots?

a) About 15 °C.   b) About 170 °F.

c) Any value between the freezing and boiling points of water.

Estimating

Converting from one type of unit to another can be done very accurately, but most of the time people use an estimate. For example, 1 inch is 2.54 cm, but people usually use 2.5 cm when they are converting from imperial to metric. In weight, 1 ounce is actually 28.3 g, but in many cookery books recipes say '1 ounce or 25 g'. They round down the 28.3 to 25, to make the calculation easier. This can make quite a difference. Using this system, 4 ounces of flour (for example) would be converted to 100 g (4 x 25 = 100). If the figures were converted accurately, 4 ounces of flour would be 113.2 g (4 x 28.3 = 113.2).

Draw the graph of this equation, and then draw the graph of each of the other equations at the top of page 40. Make sure you do this before you carry on. Mark on the gradient of the line and the y-intercept (the place where the line cuts the y axis). Compare these two numbers with the equation. What do you notice ?

Look at the gradient and the coefficient of x (the number in front of the x). They are the same. So when you look at an equation, such as $y = 3x+2$, you know immediately that the gradient of the line will be 3, because the coefficient of x is 3.

The line will also cut the y axis at +2, because the last number in the equation is +2. If the equation were $y = 4x - 5$, you would know that the gradient was 4 and the y intercept was -5.

This is why the graphs are in the form $y = mx + c$, where m is the gradient and c is the y intercept.

Make sure this makes sense to you. Try drawing some graphs of straight lines and comparing their gradient and y intercept with the equation of the line. If you are still unsure, ask your teacher for advice.

REMEMBER
Plot 3 points to check you haven't made an error. They should lie in a straight line.

Simultaneous linear equations

Look at these equations:

$y = x - 5$

$y = 16 - 2x$

These equations are both equations of straight lines, but the second equation is the equation of a line with a negative gradient. They are called simultaneous linear equations because they have to be solved at the same time. In other words, there is one value of x and one value of y for both equations. They are linear because the graphs of the equations are straight lines.

Look at the co-efficient of x. It is -2, so that means the line slopes from top left to bottom right. The first equation has a positive gradient, it slopes from bottom left to top right.

◎ *Make up a table of values, and then draw the graph. The graph should look like this:*

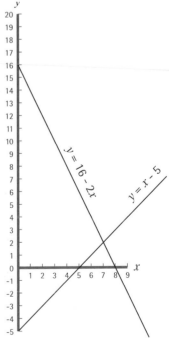

The solutions to the equations (in other words, the values of x and y) are the co-ordinates of the points where the lines cross. This point must be the solution to both equations because it is the only point that is on both lines. So $x = 7$ and $y = 2$.

The method is:

■ draw up a table of values

■ plot the graph of each equation

■ write down the co-ordinates of the point where the two lines cross.

The point where the lines cross is the solution to the equations.

BITESIZEmaths

Using graphs

Data handling

This section is about:

- calculating angles in pie charts and drawing pie charts

- correlation and drawing scatter diagrams

- the three types of average - mean, median and mode

- finding averages from grouped frequency tables

- drawing cumulative frequency curves

It is important to understand different types of data. Data is a useful tool that enables planning to take place. Imagine, for example, that there is an increase in the number of babies being born in the area where you live. The council and planning authorities can use the data about the number of babies being born to ensure that there are enough schools for the children to attend when they reach the age of five.

In this section there is some work on correlation. A correlation is a connection between two variables. It implies that they are related but it cannot be stated that they are definitely related. An example could be smoking and lung cancer.

Data handling is an area of mathematics where a number of names are used for the same thing. Scatter diagrams, scatter graphs and scattergrams are all different names for the same thing. Similarly, bar charts are also known as bar graphs.

Intermediate level students need to know how to draw cumulative frequency curves. Cumulative frequency curves are also known as ogives or S-shaped curves.

Intermediate level students also need to understand the difference between histograms and bar charts. In a histogram the frequency is proportional to the area under the curve.

The purpose of a conversion graph is to be able to read along the graph and easily convert from one unit to another. In exams, questions about conversion graphs usually involve changing money from one currency to another or changing from one unit of measurement to another, e.g. feet and inches to metres and centimetres. You can find out more about conversion graphs in the 'Using graphs' section on page 38.

FactZONE

Types of data 📺

Primary data: Data that is collected by using a questionnaire, or another method of collecting data, such as an experiment

Secondary data: Data that has been collected by someone else but that can still be used

Continuous data: Data that can take any value on a continuous scale. An example is height

Discrete data: Data that fits into groups, such as shoe size. It is possible to take a size $6\frac{1}{2}$ shoe but not a 6.999, because this size does not exist

Types of chart 📺

Pie chart: A circular chart that is split into 'slices', like a pie. Each slice is called a sector

Frequency table or tally chart: You can use this kind of table to organise or sort data

Bar chart: This is a chart where the frequency is shown by the height of the bar

A bar chart that has to show continuous data should have NO gaps between the bars.

Types of average

Mean: The arithmetic average. To work this out, add up all the data and then divide by the number of items. For example, to find the mean of 1, 2, 3, 4:

1 + 2 + 3 + 4 = 10

10 ÷ 4 = 2.5

Therefore the mean is 2.5.

Median: The middle number when the data is in order, lowest to highest or highest to lowest. In this example:

2, 3, 4, 5, **6**, 7, 8, 9,10

6 is the middle term and is therefore the median.

If there is an even number of data – e.g. 2, 3, 4, 5 – there is no middle term. Take the two middle terms and find the mean of these two:

3 + 4 = 7

7 ÷ 2 = 3.5

In this case, the median is 3.5.

Mode: The item that occurs most often. In this example:

1, 1, 2, 3, 1, 4, 5, 69, 8, 7, 5, 1, 1, 5, 1, 8

The item that occurs most often is 1. Therefore the mode is 1.

Range: The highest value minus the lowest value. It is a single value.

⊙Pie charts

Pie charts are circular diagrams – the shape of a pie – split up to show sectors, or 'slices', that represent each part of the survey. By looking at the size of each sector, it is possible to see which part of the survey had the most effect.

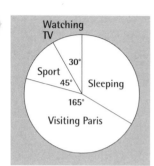

The pie chart shows how a group of students on a school trip in France spent the day. They had to travel a long way from their hostel to visit Paris.

◎ *What fraction of the day was spent sleeping?*

To find the answer, you have to work out the angle for the sleeping sector.

A pie chart is a circle and there are 360° in a circle. Add up the angles for the other sectors.

$30° + 45° + 165° = 240°$

So, the angle for the sleeping sector is:

$360° - 240° = 120°$

The fraction for sleeping is $\frac{120°}{360°}$, which cancels to $\frac{1}{3}$.

◎ *How many hours in the day were spent visiting Paris?*

The time spent visiting Paris is shown as 165° on the chart. To find the answer you first have to work out what 165° is, as a fraction of 360°.

$\frac{165}{360}$ cancels to $\frac{11}{24}$

To find out how long this is in hours, multiply the fraction by 24 hours giving an answer of 11 hours.

Calculating angles in a pie chart

Example 1

This data was collected in a survey of 300 people's favourite holiday destinations:

      60 Australia   40 UK   150 Spain   50 France

You can see that Spain is most people's favourite place, but a good way to show this data is to use a pie chart. A pie chart has more impact. It shows very clearly that in the group surveyed, Spain is a far more popular destination than the other countries.

Working out the angles

The angle at the centre of the sector is worked out in this way:

$$\frac{\text{number of people}}{\text{total in survey}} \times 360°$$

$\frac{60}{300}$ x 360° = 72° Australia $\frac{150}{300}$ x 360° = 180° Spain

$\frac{40}{300}$ x 360° = 48° UK $\frac{50}{300}$ x 360° = 60° France

Check that the angles add to 360°.

Drawing the pie chart

Draw a circle. Then draw in a radius (a line from the centre point to any point on the circumference). You can measure the angles from this point.

Example 2

The table shows the number of cars in a car park:

Type of car	Ford	Toyota	Volvo	Alfa Romeo
Number of cars	70	52	35	23

◎ *Draw a clearly labelled pie chart to show this information.*

Working out the angles

Before you begin the pie chart, you must work out the angles that are required for each sector.

Total number of cars = 70 + 52 + 35 + 23 = 180 cars

180 cars is equivalent to 360°.

Use the unitary method.

180 cars is equivalent to 360°.

So, one car = $\frac{360}{180}$ = 2°.

One car is 2° on the pie chart.

So 70 cars must be 70 x 2° = 140°.

Now you can work out the angles for the other cars and draw the pie chart.

! REMEMBER!
The angles in a pie chart must equal 360°.

Practice questions

1) The table (right) contains data showing the number of films of each category shown in a city:

Classification	18	15	12	PG	U	Total
Frequency	42	46	56	43	53	

Draw a pie chart to show this information.

2) Draw a pie chart to show this information on how people travel to school:

Transport	Car	Bus	Walk	Bike	Train
People	150	250	200	65	35

3) Draw a pie chart to show the following data:

Red balloons 30   Yellow balloons 50
Green balloons 120   White balloons 100

⊚Scatter diagrams and correlation

Scatter diagrams

You can use scatter diagrams to show correlation between two variables.

For example, do people with big hands have big feet?

◎ *What kind of data do you need to answer this question?*

You need to collect data of hand size and shoe size, and then plot it on a scatter diagram. To plot the data, draw hand size in centimetres on one axis and British shoe sizes on the other axis. Then plot the points, e.g. hand 5 cm, shoe size 6, so you plot (5,6).

Correlation

This diagram shows a correlation cloud. Here, there is a definite link and it is called a positive correlation.

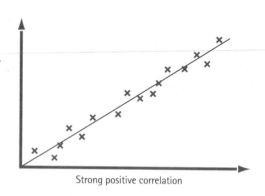

Strong positive correlation

Drawing lines of best fit

REMEMBER!
It is easier to draw lines of best fit if you use a transparent ruler.

ℹ In a case like this, the aim is to draw a line through the points so that the points are evenly distributed about the line. This means that the sum of the distances above the line is roughly equal to the sum of the distances below the line. It may mean that none of the points actually lie on the line. This is called a line of best fit.

Different types of correlation

Where the points are scattered closely around the line, there is a strong correlation.

If the points are more loosely scattered around the line, there is a moderate correlation. This means that any predictions made from the line would be rough estimates – very rough estimates if the points are widely scattered, as they are in the diagram.

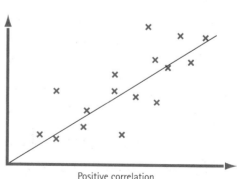

Positive correlation

If the points are so scattered that there is no obvious line, then there is no correlation.

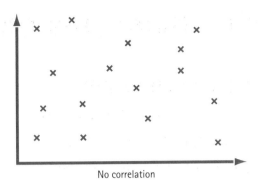

No correlation

REMEMBER!
You may be asked to 'Comment on the nature of the correlation'. This means that you have to say what type of correlation exists between the variables. Is it positive, negative, or is there no correlation?

The diagram on page 46 shows a positive correlation, where the correlation cloud is scattered from bottom left to top right. If the correlation cloud is scattered from top left to bottom right, the correlation is a negative correlation. This occurs when, as one quantity increases in a proportion, the other quantity decreases by the same proportion.

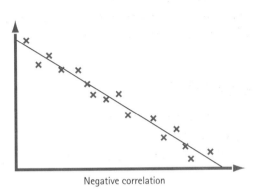

Negative correlation

Practice questions

Height (cm)	150	152	154	158	159	160	165	170	175	181
Weight (kg)	57	64	66	66	62	67	75	69	72	76

1) This table shows the heights and weights of 10 people.

a) Draw the scatter diagram.

b) 🛈 Comment on the correlation between height and weight.

c) 🛈 Nisha weighs 58 kg. Estimate her height.

On one axis, use 2 cm to represent 5 cm of height and start at 145 cm. On the other axis, use 2 cm for 5 kg, starting at 55 kg.

2) A holiday company surveyed its clients and came up with these results:

Cost of holiday	£100–£250	£250–£400	£400–£550	£550–£700	£700–£850	£850–£1000	£1000 or over
Number of clients	420 000	400 000	350 000	250 000	175 000	100 000	58 000

a) Draw a scatter diagram to show this information.

b) 🛈 Comment on the nature of the correlation, if any, between the variables.

Finding averages

The mode and the median

This frequency table shows the number of pencils that children in one class have in their pencil cases:

Number of pencils	0–4	5–9	10–14	15–19	20–24
Frequency	8	5	4	12	2

The number of pencils has been placed in groups (e.g. 0–4), so there is some information we don't know. For example, we don't know how many children had eight pencils in their pencil case, even though it may be true that more children had eight pencils than any other number of pencils. In an example like this, it is not possible to give the mode as a single figure. What is clear is that more children had between 15 and 19 pencils than any other number of pencils. This is called the modal group.

The median can also be found, by counting up the frequencies. The total is 31. Remember the median is the middle term, so count up from the left-hand side of the table. The median must be the contents of the pencil case of the sixteenth child and since that child is in the 10–14 group, then the median lies in this group.

The mean

Since we do not know exactly how many pencils each child has, it is not possible to calculate the mean accurately. But if we assume that the items in each group are evenly spread, we can use the 'half-way value' to represent the group.

Number of pencils	f	Half-way value, x	fx
0–4	8	2	16
5–9	5	7	35
10–14	4	12	48
15–19	12	17	204
20–24	2	22	44
Total	31		347

Therefore the mean number of items is approximately $347 \div 31 = 11.19$ (to 2 decimal places).

Using continuous data

For continuous data the calculation is carried out in the same way. This table gives the weight of chicks four days after hatching:

Weight, w grams	f	Half-way value, x	fx
$0 < w \leq 3$	3	1.5	4.5
$3 < w \leq 6$	4	4.5	18
$6 < w \leq 9$	12	7.5	90
$9 < w \leq 12$	6	10.5	63
Total	25		175.5

Therefore, the mean weight is approximately 175.5 ÷ 25 = 7.02 grams.

Practice questions

1) Estimate the mean value for each distribution shown below.

2) What is the modal class for each distribution?

a) 44 boxes of apples were examined and the number of damaged apples in each box was recorded:

Damaged apples	0–4	5–9	10–14	15–19	20–24
Frequency	15	10	9	6	4

b) 30 beans were prepared for cooking by rinsing in water for five days. Their roots were then measured and recorded:

Height in cm (h)	$1 \leq h < 4$	$4 \leq h < 7$	$7 \leq h < 10$	$10 \leq h < 13$
Frequency	12	8	6	4

c) 72 males were measured and their height was recorded:

Height in cm (h)	$149.5 \leq h < 154.5$	$154.5 \leq h < 159.5$	$159.5 \leq h < 164.5$	$164.5 \leq h < 169.5$	$169.5 \leq h < 174.5$
Frequency	12	15	22	14	9

d) 30 students received the following results in a test:

Mark	40 – 54	55 – 69	70 – 84	85 – 99
Frequency	4	12	9	5

e) The table shows the number of newspapers/magazines delivered to 40 houses:

Papers/Magazines	0 – 2	3 – 5	6 – 8	9 – 11
No. of Houses (Frequency)	15	7	8	10

ⓘ📺Cumulative frequency

Cumulative frequency can be thought of as a running total graph. The frequencies are added as you go along. A cumulative frequency curve is also known as an ogive.

It has this characteristic S-shape and is split into a number of parts:

• The median is at the 50% point and is known as Q2.

• The upper quartile is at the 75% point and is known as Q3 or UQ.

• The lower quartile is at the 25% point and is known as Q1 or LQ.

The upper and lower quartiles are used to find the central 50% of the distribution. This is known as the interquartile range. This is important because it shows how widely the data is spread. Half of the distribution is in the interquartile range.

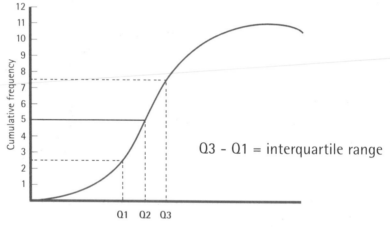

Q3 - Q1 = interquartile range

Drawing a cumulative frequency diagram

This table illustrates the age distribution in a village of 720 people:

Age	0–9	10–19	20–29	30–39	40–49	50–59	60–69	70–79	80–100
Frequency	48	72	65	120	153	50	72	96	44

We can draw a cumulative frequency diagram to illustrate this data.

Age (less than)	10	20	30	40	50	60	70	80	80–100
Cumulative frequency	48	120	185	305	458	508	580	676	720

The cumulative frequency table is built up by adding the frequencies to what came before, so the 120 in the third cell is 48 + 72 (from the first table). The 185 in the fourth cell is the 120 from the third cell + 65 from the first table, and so on.

❗ REMEMBER! Age 0-9 means '0-9 years 364 days', that is, nearly 10. Age is a difficult variable – you must be careful in the exam.

◎ *Draw the curve by plotting age against cumulative frequency.*

We have included a sketch of how it should look, but your diagram should be a clear graph showing the ogive.

Note that the points are plotted using the upper boundary of the class. Mark on the Q1, Q2 and Q3 as we have done and notice how they go across to the curve and then down to the age-axis. The value of Q1 is the value the downward line makes with the age-axis, and not 25%.

◎ *Use your graph to find the median and the interquartile range.*

Remember that the median is Q2, so draw a line from the 50% mark on the cumulative frequency axis, i.e. half-way through the distribution, across to the curve and then down to the age-axis. This value is the median.

◎ *Find the interquartile range.*

To do this, draw a line from Q3 and Q1 across to the curve and then down to the age-axis.

The interquartile range = Q3 - Q1.

Make sure you take your estimate from your own graph. Q3 on this graph is 64, Q1 is 30 , so the interquartile range is 64 - 30 = 34. Your estimate should be close to this answer, if not the same.

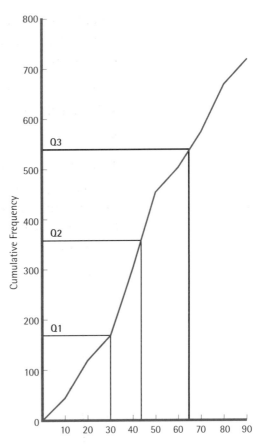

Practice questions

1) A biologist measures the lengths of worms needed to make compost. Copy and complete the table, then draw the cumulative frequency curve for the results below:

Length (*cm*)	Frequency	Cumulative frequency	Upper limit
0–10	150	150	≤10
10–20	132	282	≤ 20
20–30	126		
30–40	75		
40–50	9		

2) This is the age distribution of the population of a country:

Age	Number of people (Millions)
under 10	16
10-19	12
20-29	17
30-39	16
40-49	15
50-69	10
70-89	4

Use the information in the table above to:

a) draw a cumulative frequency table

b) draw the cumulative frequency curve for this information.

Algebra

This section is about:

- number patterns, sequences and making predictions

- finding a rule between two variables in an investigation

- multiples, factors and prime numbers

- solving equations with unknown numbers on one or both sides

- expanding brackets in an algebraic expression

- solving equations with fractions

- factorising algebraic terms

- changing the subject of a formula

- solving simultaneous linear equations

Many students develop a fear of algebra because they find it hard. If you get stuck, think about what you would do if you were dealing with numbers instead of letters. Algebra is based on the same set of rules, so you can use the same methods to solve problems. It is important to work through each topic in this section, in sequence. It will help you to understand the work.

Spotting patterns is important. It means that a mathematician can make predictions about what is going to happen. Some numbers have special names because of their patterns. Square numbers make square patterns, cube numbers make cube patterns, and triangular numbers make triangular patterns. There is some work on investigating patterns and finding connections between variables in this section.

There is also some work on factors and prime numbers, and on writing a number as a product of its prime factors.

When you are working on equations, remember that they need to be treated as a balance. Whatever you do to one side, you must do to the other side too.

The emphasis in this chapter and in the TV series is on techniques. Here, we have aimed to cover most of the basic techniques but there is a limited amount of space.

Make sure you understand how to use directed numbers (see Factzone page 67) Most students cover this in Year 9 or earlier, so you may need to look back over this work.

Intermediate-level students need to practise changing the subject of an equation. The technique of multiplying both sides of an equation by -1 can be very useful.

There are some exam-style questions at the end of the section. Use these to practise your techniques, and remember that you will have to draw on knowledge from other areas of mathematics to solve problems.

Multiples, factors and prime numbers

Multiple: The answer after multiplying a certain number, e.g. the multiples of 5 are 5, 10, 15, 20, etc.

Factor: A number that divides exactly into another number with no remainder, e.g. the factors of 8 are 1, 2, 4, 8, because they are the numbers that divide exactly into 8.

Prime number: This is a number with only two distinct factors, e.g. 5 is a prime number because it can only be divided by 1 and 5.

Sequence of prime numbers: 2, 3, 5, 7, 11, 13, 17, 19, etc.

Prime factor: this is a factor that is also prime. Exam questions ask for a number to be written as a product of its prime factors.
For example, 20 written as a product of its prime factors is 2 x 2 x 5.

Shorthand mathematics

Squared means multiplied by itself so 3^2 means 3 x 3 = 9.

Cubed means multiplied by itself and then multiplied by itself,

e.g. 4^3 means 4 x 4 = 16, 16 x 4 = 64,
so, 4^3 = 64

Square root: The opposite to squaring, $\sqrt{}$ is the symbol for square root.

$\sqrt{9} = 3$ because $3^2 = 9$
$\sqrt{100} = 10$ because $10^2 = 100$

Squares	Square roots	Cubes	Cube roots
$1^2 = 1$	$\sqrt{1} = 1$	$1^3 = 1$	$\sqrt[3]{1} = 1$
$2^2 = 4$	$\sqrt{4} = 2$	$2^3 = 8$	$\sqrt[3]{8} = 2$
$3^2 = 9$	$\sqrt{9} = 3$	$3^3 = 27$	$\sqrt[3]{27} = 3$
$4^2 = 16$	$\sqrt{16} = 4$	$4^3 = 64$	$\sqrt[3]{64} = 4$
$5^2 = 25$	$\sqrt{25} = 5$	$5^3 = 125$	$\sqrt[3]{125} = 5$

Algebra

4 times x is written as $4x$, so miss out the times sign.

Variables

This is the name given to values that change, usually the letters in algebra, for example, x or y in an equation.

Collecting like terms

You can only add together terms that are alike. In algebra this generally means the same letter:

$a + a + a = 3a$, so $a + a + a + b = 3a + b$

Note

$a + a + b = 2a + b$

You cannot add as and bs together.

$a^2 + a^2 + a = 2a^2 + a$

a^2 and a are not like terms, so they cannot be added.

Multiplying and dividing like terms

It is easier if you group the numbers, then the letters in alphabetical order,

e.g. $5y$ x $-6x$ x $3z$ = 5 x -6 x 3 x x x y x z
= $-90xyz$

Substitution

In algebra, substitution means that a letter is taken out and replaced by a number. The numbers are then used to find the numerical value for the algebraic expression,

e.g. if $x = 2$ and $y = 3$, $xy = 2$ x $3 = 6$

Number patterns

Square numbers

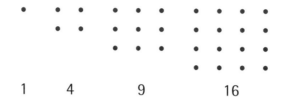

1   4   9   16

Each of the numbers in this sequence can be shown as a square pattern, so they are called square numbers.

◎   *1, 4, 9, 16, 25, 36, 49. Fill in the missing numbers in this sequence.*

To answer this question, you may need to rewrite the numbers like this:

    1 x 1   2 x 2   3 x 3   4 x 4

Now it is easier to see that the next one must be 5 x 5, then 6 x 6 and so on. This is called the sequence of square numbers.

! REMEMBER
You must be able to recognise this sequence. A question like this often appears on exam papers.

Triangular numbers

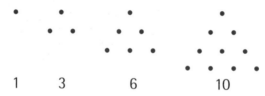

1   3   6   10

(?)   *What are the next three numbers in the sequence?*

Look at the pattern again. It starts with 1, then it is 1 + 2, then 1 + 2 + 3, then 1 + 2 + 3 + 4. The next number will be the answer to 1 + 2 + 3 + 4 + 5, which is 15.

Look at the shape of the pattern. It is a triangle, so these numbers are called triangular numbers.

There is another way of showing the pattern. It is still triangular but it looks like this:

! REMEMBER
You should know both arrangements, for the exam.

Cube numbers

Look at each of these diagrams. They are cube numbers, so the first cube number is 1.

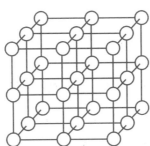

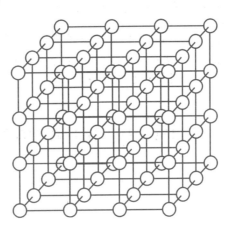

In mathematical shorthand:

$1^3 = 1 \times 1 \times 1 = 1$

$2^3 = 2 \times 2 \times 2 = 8$ (Don't confuse this with $2 \times 3 = 6$, this is $2 \times 2 = 4$, then $4 \times 2 = 8$)

$3^3 = 3 \times 3 \times 3 = 27$ ($3 \times 3 = 9$, then $9 \times 3 = 27$)

$4^3 = 4 \times 4 \times 4 = 64$ ($4 \times 4 = 16$, then $16 \times 4 = 64$)

Practice questions

1) Explain what square numbers and triangular numbers are and how they look.

2) 1, 8, 27, 64 ,___, ___

a) What is this special sequence called?

b) Write an explanation of how it is built up.

c) Write down the next two numbers in the sequence.

3) 1, 3, 6, 10, ___, ___, ___

a) Write down the next three terms in this special sequence.

b) Show the sequence as a series of diagrams.

c) What name is given to this sequence?

What is a sequence ?

A sequence is a set of numbers that are connected in some way. For instance, a very simple sequence is 3, 6, 9, 12.

Obviously this is the three times table, otherwise known as the multiples of three. Most students recognise this sequence and simply add the next three numbers: 15, 18, 21.

REMEMBER
A sequence is shown as 1, 4, 9, 16, 25, __, __, __,
(where __, __, __, means continue in the same way).

Example 1

1, 4, 9, 16, 25,__, __, __, __, __, __

◎ *Write down the next six numbers in this sequence. Explain how it is made.*

Do this before you read any further.

This sequence is very important and you need to remember it. It is called the sequence of square numbers.

Let's look at the sequence again:

1, 4, 9, 16, 25, __, __, __

Another way of writing this sequence is:

1 x 1   2 x 2   3 x 3   4 x 4   5 x 5   and so on.

Now it is easy to see how the sequence actually generates.

Example 2

1, 1, 2, 3, 5, 8, __, __, __

◎ *Can you see how the sequence is building up?*

This is a very famous sequence, discovered by a medieval Italian mathematician called Fibonacci. It is called the 'Fibonacci sequence'.

The pattern builds up by adding together the previous two numbers.

It starts at 1. There is nothing in front of the first one, so 0 + 1 = 1. We generate the second 1, 1 + 1 = 2 . Now we generate the 2. In the same way, carrying on like this:

1 + 2 = 3, 2 + 3 = 5, 3 + 5 = 8 and so on.

◎ *Write out the first 10 numbers in the Fibonacci sequence.*

56

Practice questions

For questions 1-6 below, find the next number in each sequence:

1) 2, 4, 8, 16

2) 13, 18, 24, 31

3) 400, 200, 100, 50

4) 1, 3, 9, 27

5) 3800, 380, 38, 3.8, 0.38

6) 2, 10, 50, 250

7) The rule below is used to work out each number from the next integer starting at 3.

Multiply by 3 and then add 1.

Starting off with 3 we get 10, 13, 16.

a) Write down the next number.

b) Using the same rule but a different starting number, the second number is 25. What was the start number?

c) Aled starts with 3, then 4 and then 5 and uses a different rule to generate the numbers 17, 22, 27.

Fill in the box to complete the new rule.

8) To generate a sequence of numbers, Hefin multiplies the previous number in the sequence by 3, then adds 1.

Here are the first 3 numbers of his sequence.

4, 13, 40, __

a) What is the next number in the sequence?

b) These are the first 4 numbers in the Fibonacci sequence:

1, 1, 2, 3, __, __

Find the next 2 numbers in the sequence.

9) What is the next number in this sequence:

2, 5, 11, 23, __ ?

The Coronet skaters, in the TV programme, showed how they can make a star pattern. Look at how they built up the star.

The girls started by placing one skater at the centre. Then a skater joined on each side. Look at the pattern. (Always write the number of skaters under each diagram.)

5

Then an extra skater joins on.

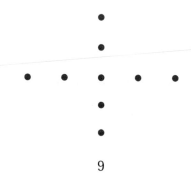

9

They continue to build up the pattern by adding a skater on each side.

We can build up a table showing the pattern number and the number of skaters.

Pattern number (p)	Number of skaters (s)	Differences
1	1	
		4
2	5	
		4
3	9	
		4
4	13	

Finding a rule

Look at the 'Differences' column in the table above. These are the differences in the number of skaters, so in other words, 4 skaters are added each time.

This is important. It tells us that the rule or equation that links the pattern number and the number of skaters is a linear equation, because the differences are all the same. This means that the equation does not contain powers of p greater than 1.

The differences are constant and they are 4, so multiply the pattern number by 4:

$4 \times 1 = 4, 4 \times 2 = 8$

(?) *Does this give the number of skaters? Try some for yourself and make sure you are clear about what is going on.*

Multiplying by 4 is not enough. The pattern number (p) x 4 does not give the number of skaters. For example, $4 \times 1 = 4$, not 5. We need to do something else as well. We need to subtract 3.

The rule is **Pattern number x 4 - 3 = number of skaters**.

In the exam you should try to write the rule using algebra.

In algebra the rule reads as $4p - 3 = s$

For example, if $p = 1$:   if $p = 2$,

$$s = 4 \times 1 - 3 \qquad\qquad s = 4 \times 2 - 3$$
$$= 4 - 3 \qquad\qquad\qquad = 8 - 3$$
$$= 1 \text{ which is correct} \qquad = 5$$

You can check this by looking back at the diagrams.

Now we can predict the number of skaters for pattern number 5:

$$s = 4 \times 5 - 3$$
$$= 20 - 3$$
$$= 17$$

We can check this by drawing the diagram:

Why is it + 4?

One skater is added on each of the 4 arms of the diagram, i.e. there are 4 more each time.

Investigating patterns 2

Here, we have patterns made up of matches. Again we can make a table of values and look at the differences.

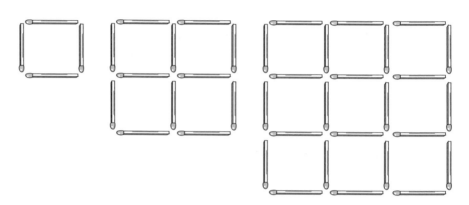

| | 1 x 1 | 2 x 2 | | 3 x 3 |

Pattern number (n)	Number of matches (m)	First differences
1	4	
		8
2	12	
		12
3	24	
		16
4	40	
		20
5	60	
		?
6	?	

◎ *Copy and complete the table. Use the differences to predict the number of matches for pattern 6. Draw pattern 6 to check if your prediction was right.*

Let's look at the differences again.

    8, 12, 16, 20, __

Obviously the next in the pattern is 24.

⊙ *Check this and make sure that you agree it is correct.*

As the first differences are not the same, look at the second differences.

	8		12		16		20		24
		4		4		4		4	

REMEMBER There are usually some questions like this in the exam. Practise making up tables of differences for investigations, looking at the first or the second differences, and using the clues that this gives you, to find the rule.

Here the second differences are all the same, a multiple of two. This is a clue to the rule for this investigation. The rule must be a quadratic, which means it has a squared term in it.

To find the rule, look at this table.

n	m	n^2	$2n^2$	
1	4	1	2	This line shows that there are 4 matches in pattern 1. If we square the pattern number we get 1. If we then double this number we get 2.
2	12	4	8	In pattern 2, there are 12 matches. $2^2 = 4$, $2 \times 4 = 8$
3	24	9	18	In pattern 3, there are 24 matches. $3^2 = 9$, $2 \times 9 = 18$
4	40	16	32	
5	60	25	50	

If you look at the column headed $2n^2$, you will see that we have to add a multiple of 2 to the numbers in this column to get the number of matches.

On the first line, the number in the $2n^2$ column is 2. If we add 2×1 to this, we get 4, the number in the matches column (m).

On the second line, the number in the $2n^2$ column is 8. Add 2×2 and we get 12, the number of matches, and so on.

This means that the rule connecting the number of matches and the diagram number must be $m = 2n^2 + 2n$, where m is the number of matches and n is the pattern number.

So for pattern 3, using the rule $m = 2n^2 + 2n$:

the number of matches $= 2(3)^2 + 6$

$= 18 + 6$

$= 24$ matches

If you look at the table, you will see that pattern 3 does indeed use 24 matches.

Multiples, factors and prime numbers

Many students get confused by these types of number and find it difficult to remember which is which. That's why they are in the Factzone and here.

A **multiple** is the answer after multiplying by a certain number, e.g. the multiples of 5 are the answers in the 5 times table.

The multiples of 5, up to 50, are 5, 10, 15, 20, 25, 30, 35, 40, 45 and 50.

The multiples of 3, up to 15, are 3, 6, 9, 12, 15.

❗ R E M E M B E R Intermediate and higher level students need to know that factors are also used in factorising algebraic expressions.

A **factor** is a whole number that divides exactly into another number with no remainder.

So the factors of 8 are 1, 2, 4 and 8, because all those numbers divide exactly into 8 with no remainder.

A **prime number** is a number with only 2 factors and those factors are different. The first prime number is 2, because its factors are 1 and 2. In other words the only numbers that go exactly into 2 are 1 and 2. The next prime number is 3, because it only has 2 factors, 1 and 3, and they are different. 1 is not a prime number, as it does not have two factors that are different.

Multiples and factors

Nisha is thinking of a number.

Look at what each cloud is telling us.

◎ *Write the multiples of 3 that are less than 20 and the multiples of 5 that are less than 20.*

Whatever the number is, it must be in both lists. The only number that is in both lists is 15, so Nisha must have been thinking about 15.

It is a multiple of 3.

My number is less than 20.

It is a multiple of 5.

Prime numbers

◎ *Think of the numbers more than 1 but less than 10. Which of them are prime numbers?*

Step 1: Write the numbers down first: 2, 3, 4, 5, 6, 7, 8, 9

Step 2: Look at each number in turn and decide what numbers divide exactly into that number

Number	Factors				Prime? Y/N
2	1	2			Y
3	1	3			Y
4	1	2	4		N
5	1	5			Y
6	1	2	3	6	N
7	1	7			Y
8	1	2	4	8	N
9	1	3	9		N

So the prime numbers less than 10 are 2, 3, 5, 7. They are all the numbers with two different factors. Numbers with more than two factors are not prime.

> **! REMEMBER**
> You have to be able to spot the sequence of prime numbers in the exam. You might get a question that asks you, 2, 3, 5, 7, 11, __, __, __. Fill in the three missing numbers in this sequence.

Algebra

Prime factors

You might be asked to write a number as a product of its prime factors. This example uses 20.

Prime factor	Number	
2	20	Put the number 20 in the right column of the ladder and the first prime factor, 2, in the left column. Divide 20 by 2 and put the answer (10) under the 20.
2	10	Will the chosen prime factor (2) still divide into this number? If yes, proceed as above.
		Will the chosen prime factor divide into this number? If no, pick the next prime factor, in this case 3.
		Will 3 divide into this number? If yes, use it. If no, discard it. In this case the answer is no, so we need to discard it and pick the next prime factor after this number, which is 5.
5	5	You know when you have finished because you have a 1 in the last cell on the right-hand side of the ladder.
	1	

To write the number as a product of its prime factors, read down the left column. In this case 20 as a product of its prime factor is 2 x 2 x 5. To test this answer, multiply the numbers out and you should get the start number.

Solving equations with an unknown number on one side

Example 1

This is a straightforward equation that should make the method of solving it clear:

$x + 7 = 10$

This means a number + 7 = 10. The missing number must be 3, so $x = 3$.

If we didn't know the missing number, the way to work it out would be to treat the equation like a balance. Everything that is done to one side must also be done to the other side.

In this example, the 7 is taken away from the left-hand side of the equation to leave x on its own:

$x + 7 - 7 \quad = x$

You must also take 7 from the right-hand side of the equation, to balance it:

$x + 7 - 7 \quad = 10 - 7$

so, $x \quad\quad = 3$

Example 2

Solve: $4x + 14 = 26$

$4x + 14 \quad\quad = 26$ First of all take 14 from both sides.

$4x + 14 - 14 = 26 - 14$

$4x \quad\quad\quad = 12$ This means that four times the number we want = 12, but we don't want four times the number, we want the number itself. We need to divide both sides by 4.

$\frac{4x}{4} \quad\quad\quad = \frac{12}{4}$ The 4s on the left-hand side of the equation cancel out.

$x \quad\quad\quad = 3$

(?)   *How could you check the answer is correct?*

To check your answer, put the value of x into the original equation. Here:

$4 \times 3 + 14 = 26$

so the answer is correct.

Example 3

Solve: $5x - 4 = 26$

$5x - 4$	$= 26$	Remove the -4 on the left-hand side by adding 4. Now, 4 has to be added to the right-hand side as well.
$5x - 4 + 4$	$= 26 + 4$	
$5x$	$= 30$	Next, divide both sides by 5.
$\frac{5x}{5}$	$= \frac{30}{5}$	The 5s on the left-hand side cancel out.
x	$= 6$	

Example 4

Solve: $9 = 3y + 15$

This equation is written the other way round, and you have to find y instead of x. Don't panic. You still need to treat this as a balance, in the same way as the other examples. The y is just a different way of writing an unknown number.

9	$= 3y + 15$	Subtract 15 from both sides.
$9 - 15$	$= 3y + 15 - 15$	
-6	$= 3y$	Divide both sides by 3.
$\frac{-6}{3}$	$= \frac{3y}{3}$	The 3s on the right-hand side cancel out.
		Divide -6 by 3 to get -2.
y	$= -2$	

Be careful when dividing the minus number on the left-hand side. It is easy to get confused with this type of question. Ask yourself 'What do I have to multiply 3 by to make -6?'

> **REMEMBER**
> Align the equal signs in your working. It makes your answer look neater and increases your chance of higher marks.

Practice questions

Solve these equations:

1) $x + 3 = 7$

2) $x + 11 = 21$

3) $x + 10 = 30$

4) $x + 40 = 25$

5) $x - 6 = -3$

6) $y - 8 = -10$

7) $8 + y = 17$

8) $w - 19 = 41$

9) $5 + x = 12$

10) $4x + 1 = 17$

11) $5x - 4 = 16$

12) $9x - 7 = 11$

13) $4 + 4x = 40$

14) $50y - 1 = 49$

15) $13 = 4x - 7$

16) $56 = 8x - 8$

17) $200y - 50 = 350$

18) $3 + 14x = 45$

19) $60 = 42 + 9y$

20) $0 = 4x - 1$

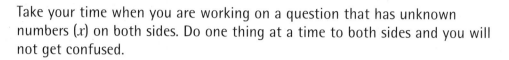

Solving equations with unknown numbers on both sides

66

Take your time when you are working on a question that has unknown numbers (x) on both sides. Do one thing at a time to both sides and you will not get confused.

Example 1

Solve: $8x - 4 = 6x + 2$.

$8x - 4$	$= 6x + 2$	Here, add 4 to both sides. This will remove the -4 from the left-hand side of the equation.
$8x - 4 + 4$	$= 6x + 2 + 4$	
$8x$	$= 6x + 6$	Take $6x$ away from both sides. This will remove the $6x$ from the right-hand side.
$8x - 6x$	$= 6x + 6 - 6x$	
$2x$	$= 6$	Divide both sides by 2.
$\frac{2x}{2}$	$= \frac{6}{2}$	The 2s on the left-hand side cancel.
x	$= 3$	

Example 2

Solve: $3x + 8 = 40 - 5x$

$3x + 8$	$= 40 - 5x$	Remove the 8 from the left-hand side by subtracting 8 from both sides.
$3x + 8 - 8$	$= 40 - 5x - 8$	The 8 on the left-hand side will disappear and we can tidy up the right-hand side.
$3x$	$= 32 - 5x$	Now add $5x$ to both sides. This will remove the $-5x$ on the right-hand side of the equation.
$3x + 5x$	$= 32 - 5x + 5x$	
$8x$	$= 32$	Now divide both sides by 8.
$\frac{8x}{8}$	$= \frac{32}{8}$	The 8s on the left-hand side cancel.
x	$= 4$	

Practice questions

Solve these equations.

1) $6x - 3 = 4x + 2$

2) $5x + 4 = 6x - 5$

3) $7x - 7 = 6x + 7$

4) $11x + 6 = 5x + 30$

5) $5x + 1 = 8 - 2x$

6) $9x - 3 = 6x + 6$

7) $10y + 2 = 12y - 18$

8) $12x - 12 = 6x + 6$

9) $2x + 15 = 4x + 3$

BITESIZEmaths

Brackets and factorising

Remember these are the opposite of each other. When expanding brackets, multiply everything inside the bracket by what sits outside the bracket. For example:

$3x(x + y) = 3x^2 + 3xy$

Directed numbers

-4 is 4 less than 0, e.g. -4°C is 4 degrees less than 0.

-4 + 4 is 4 less than 0 plus 4, so the answer is 0.

4 - -4 = +4 + 4 = 8 The two minus signs in the middle become a plus.

-4   -3   -2   -1   0   1   2   3   4

Directed numbers in algebra

These work in the same way as numbers. For example:

$-w + w = 0$

$y - -y = y + y = 2y$

Highest common factors

The highest common factor is the highest number that goes into two numbers. For example, the highest common factor of 12 and 15 is 3. To find the highest common factor of two numbers, write down all the factors of the two numbers and look for the highest number that appears in both lists.

Quadratics ⓘ

These are terms that contain a power of 2. For more information on quadratics, visit our on-line service at http// www.bbc.co.uk

Straight lines ⓘ

Straight lines form the general equation $y = mx + c$, where m is the gradient of the line and c is the y-intercept (the place on the y-axis where the graph cuts the axis).

This is called the general form of a linear equation. In linear equations, the highest power is 1. In other words, there are no powers of 2 or more.

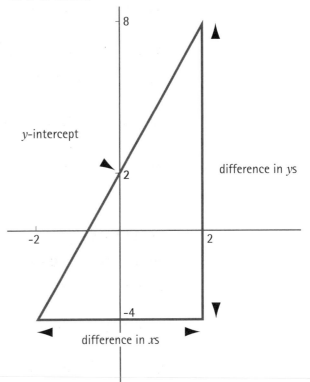

In the sketch, you can see the line cuts the y-axis at +2, so this is the last part of the equation.

The gradient = $\dfrac{\text{difference in } ys}{\text{difference in } xs}$

In this case the gradient is 3, so the equation is $y = 3x + 2$.

Just substitute the numbers from the sketch into the equation $y = mx + c$.

You'll find more information in the TV series.

As you probably know, when you are dealing with numbers, you must work out the bracket first. For example, if you were asked to work out 7 = (8 - 1), you would have to work out the bracket first.

In algebra, it is different.

$2(x + 1)$, for example, is a product. It means $2 \times (x + 1)$. Each term inside the bracket needs to be multiplied in turn by what is outside the bracket. So in this case:

$2 \times x$ is $2x$ and 2×1 is 2

so, $2(x + 1) = 2x + 2$

Example 1

Solve: $3(x - 1) = 2(x + 2)$

$3(x - 1) = 2(x + 2)$	Expand both sides by multiplying out what is inside the brackets.
$3 \times x = 3x$, then $3 \times -1 = -3$	Expand the left-hand side first, so that the left-hand side becomes $3x - 3$.
$2 \times x = 2x$, then $2 \times 2 = 4$	Then expand the right-hand side, so that it becomes $2x + 4$. Now we can rewrite the whole equation and solve it using the method we used on page 66.

◎ *Try to solve the equation.*

$3x - 3 = 2x + 4$

$3x - 3 + 3 = 2x + 4 + 3$

$3x = 2x + 7$

$3x - 2x = 2x + 7 - 2x$

$x = 7$

❓ *How can you check the answer?*

Check your answer by putting the value for x, which is 7, into the original equation:

Left-hand side: $3(7 - 1) = 3 \times 6 = 18$

Right-hand side: $2(7 + 2) = 2 \times 9 = 18$

❗ **REMEMBER**
Treat the equation like a balance. Do the same thing to both sides of the equation.

Example 2

Solve: $6(2x + 1) = 2(x + 1) + 24$

$6(2x + 1)$	$= 2(x + 1) + 24$	As in the previous example, multiply out the brackets first.
$12x + 6$	$= 2x + 2 + 24$	Now tidy up the right-hand side.
$12x + 6$	$= 2x + 26$	Now we can use the method for solving equations with x on both sides.
$12x + 6 - 6$	$= 2x + 26 - 6$	
$12x$	$= 2x + 20$	
$12x - 2x$	$= 2x + 20 - 2x$	
$10x$	$= 20$	
$\frac{10x}{10}$	$= \frac{20}{10}$	
x	$= 2$	

◎ *Check this answer by substituting 2 for x in the original equation.*

Practice questions

All students need to be able to solve equations with brackets, but Foundation-level students might find some of these questions quite difficult.

Solve these equations.

1) $2(x + 1) = 3x - 8$

2) $4(x - 3) = 5(x - 5) + 5$

3) $10(x - 3) = 6x - 2$

4) $9(x - 8) = 12x - 12$

5) $3x + 2(x + 1) = 2x + 23$

6) $5(x + 2) + 3 = 6x - 4$

7) $4(3x + 2) + 2(x + 1) = 52$

8) $5(x - 3) + 3(x + 4) = 7x + 4$

9) $3(2x + 4) + 4 - 4(x + 7) = 0$

10) $4(x - 1) + 5(x + 3) = 2(5x + 3)$

11) $5(x - 4) + 4x = 3x - 2$

12) $6x - 2(x - 1) = 3x + 5$

13) $6(x + 2) = 5x + 20$

14) $3(4x + 3) - 2(3x - 5) = 15x + 1$

15) $4(x + 2) + 3x = 78$

16) $5(y - 3) - 2y = 2y - 3$

17) $2(2z + 7) = 5z - 1$

18) $4(x + 3) - 2(x - 5) = 2(3x + 1)$

19) $14(m + 3) = 35m$

20) $5(x - 4) = x + 4$

Solving equations with fractions

Look at what happens when fractions are multiplied:

Clearly $2 \times \frac{1}{2} = 1$

Look at what happens with the numbers. We could just as easily have cancelled the 2s, leaving 1 as the answer. This works with any fraction. Try it yourself for $3 \times \frac{1}{3}$, $4 \times \frac{1}{4}$, and so on.

(?) *Why do you think this works?*

The number that is being used to multiply the fraction is the same number as the denominator (the bottom part of the fraction). So the fraction can be cancelled. It works the same way when you are using fractions in algebra.

! REMEMBER
In the exam, you might have to find the value of y or another letter, instead of x. It makes no difference to the way you solve the equation.

Example 1

Solve: $\frac{6}{x} = 12$

In a question like this, the first thing you need to do is get the x out of the denominator and make it a numerator (the top part of a fraction). This makes it easier to deal with later in the solution. To do this, multiply both sides by x. It works in the same way as the numbers in the fractions we tried above.

$x \times \frac{6}{x}$	$= 12 \times x$	The xs on the left-hand side of the equation cancel out.
6	$= 12x$	Now we have an equation that looks like the ones we worked on earlier.
$\frac{6}{12}$	$= \frac{12x}{12}$	Remember, the 12s on the right-hand side cancel.
$\frac{1}{2}$	$= x$	Remember $\frac{6}{12} = \frac{1}{2}$.

Example 2

Solve: $\frac{2x}{3} = 5$

Here, x is already part of the numerator, but we can use the same method as in example 1.

$\frac{2x}{3}$	$= 5$	Multiply both sides by 3 to remove the 3 on the left-hand side of the equation.
$3 \times \frac{2x}{3}$	$= 5 \times 3$	The 3s on the left-hand side of the equation then cancel. We are left with an equation that looks like those we have worked on before.
$2x$	$= 15$	Now divide both sides by the coefficient of x (the number in front of the x). Here, we need to divide by 2.
$\frac{2x}{2}$	$= \frac{15}{2}$	The 2s on the left-hand side cancel.
x	$= 7.5$	

! REMEMBER
Not all equations work out to be exact whole numbers.

Practice questions

Solve these equations.

1) $\frac{3}{x} = 6$

2) $\frac{4}{x} = 7$

3) $\frac{5}{x} = 2$

4) $\frac{12}{x} = 8$

5) $\frac{7}{y} = 3$

6) $\frac{9}{t} = 12$

7) $\frac{15}{m} = 10$

8) $\frac{5}{a} = 20$

9) $\frac{2}{h} = 1$

10) $\frac{6}{x} = 9$

11) $\frac{x}{2} = 19$

12) $\frac{t}{4} = 7$

13) $\frac{m}{8} = 16$

14) $\frac{f}{3} = 12$

15) $\frac{z}{15} = 11$

16) $\frac{6n}{7} = 13$

17) $\frac{3x}{4} = 10$

18) $\frac{9x}{10} - 3 = 24$

19) $\frac{4z}{7} = 12$

20) $\frac{5x}{12} = 6$

Practice questions

Use what you know about solving equations to solve these problems.

$y + 4$

$y - 2$

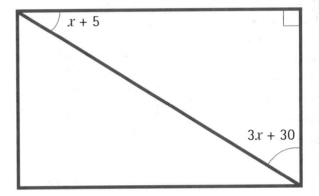

1) Find y if the perimeter is 32 cm.

2) The length of a rectangle is 5 cm more than its width. If the perimeter is 46 cm, find the width.

3) Find the angles in the triangle above.

4) The sum of three consecutive numbers is 459. Find the numbers.

5) The angles of a triangle are 23°, $y°$ and $(5y + 7)°$. Find the value of y.

ⓘ 📺 Factorising

In algebra, factorising is the opposite of expanding brackets. Remember, a factor is a number that divides into another number. When you are factorising, look for terms that divide exactly into the term you want to factorise.

Example 1

❓ *Look at this expression: 14a + 4b. What is the highest term that will go into both 14 and 4?*

It is 2, so we need to take out the 2.

2(   )   Now we need to decide what to multiply the 2 by to get $14a$. Multiplying 2 by $7a$ gives us $14a$, so the first term in the bracket is $7a$.

$2(7a$)   Now we need to ask the same question for the $+ 4b$. Multiplying 2 by $2b$ gives us $4b$.

❓ *Can you work out how to write the answer?*

The answer is: $2(7a + 2b)$.

You can check this by expanding the brackets. This should take you back to $14a + 4b$.

Example 2

Factorise: $4x^2 - 5x$

$4x^2 - 5x$ Here, we need to identify the common factor in both terms. It is x, so we can take out the x and work out what it needs to be multiplied by, to make the terms in the brackets.

$x(4x - 5)$ This one is quite straightforward. This is the final answer.

Example 3

Factorise: $4ab + 6b^2$

$4ab + 6b^2$ Here the common factors are in the numbers and the letters.

We need to take out 2, because 2 is a factor of both 4 and 6. The letters also have a common factor: b is a factor of ab and b^2.

$2b($)   Now we have to decide what to multiply $2b$ by to make $4ab$ and $6b^2$. The first term in the brackets must be $2a$, because $2a \times 2b$ will give us the $4ab$.

$2b(2a$)   The second term must be $+3b$, because $2b \times +3b$ gives $+6b^2$

$2b(2a + 3b)$ This is the final answer.

The difference of two squares

There is a special case in factorising, which is known as 'the difference of two squares'. This refers to expressions such as: $x^2 - 4$. Both the terms are squares. Therefore both have a square root. In mathematics 'difference' means subtract. The difference between two squares is one square term minus a second square term.

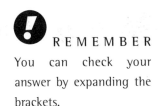
REMEMBER
You can check your answer by expanding the brackets.

73

Algebra

Example 1

If we factorise $x^2 - 4$, we get $(x - 2)(x + 2)$. But why is this so?

(?) *If you expand $(x - 2)(x + 2)$, what happens to the middle term?*

You get $x^2 + 2x - 2x - 4$.

Notice how the two middle terms cancel each other out, leaving you with $x^2 - 4$.

Example 2

Factorise $16x^2 - 81$

Again, this is the difference of two squares.

It factorises to $(4x + 9)(4x - 9)$.

Practice questions

Complete these statements.

1) $2x + 4y = 2(x + \quad)$

2) $9b + 15c = 3(\quad + 5c)$

3) $12a + 6b = 6(\quad + b)$

4) $4m + 8n = 4(\quad + \quad)$

5) $15b + 20c = 5(\quad + \quad)$

6) $9v - 12w = 3(\quad - \quad)$

7) $6v - 12u = 6(\quad - \quad)$

8) $25r + 30s = \quad(5r + \quad)$

9) $24t + 36r = \quad(2t + \quad)$

10) $27m - 81n = \quad(m - \quad)$

Factorise these expressions.

11) $12a + 8b$

12) $15x - 24y$

13) $54c - 27d$

14) $35m - 28t$

15) $12k - 16l$

16) $40h - 24j$

17) $15a - 12b - 6c$

18) $8x^2 + 16xy + 24y$

19) $15x^2 - 35y^2$

20) $64p^2 - 32p$

Decide if these expressions are the difference of two squares. If they are, factorise them as explained above.

21) $x^2 - y^2$

22) $4m^2 + 9y^2$

23) $49t^2 - 64m^2$

24) $9z - 16y$

25) $12m^2 - 16n^2$

26) $25h^2 + 36j^2$

27) $100k^2 - 81m^2$

28) $169c^2 - 225d^2$

Changing the subject of a formula

Changing the subject of a formula means rearranging an equation or formula to get one letter on its own and all of the other letters on to the other side of the equation. It is also known as transformation of formulae.

Example 1

Make x the subject of $x + a = c$.

This means get x on its own and everything else on to the other side of the equation.

$x + a = c$ means 'a number plus another number gives the third number'. You have to get x on its own.

If this were three numbers, say $3 + 2 = 5$, to get 3 on its own you would take the 2 away. To balance the equation you would need to take 2 away from the right-hand side as well. Use the same method in algebra.

$x + a$	$= c$	Take a away from both sides.
$x + a - a$	$= c - a$	The as on the left-hand side subtract to equal nothing.
x	$= c - a$	

Example 2

Make b the subject of $a(a + b) = c^2$

$a(a + b)$	$= c^2$	First, expand the brackets.
$a^2 + ab$	$= c^2$	Now subtract a^2 from both sides.
$a^2 + ab - a^2$	$= c^2 - a^2$	Remove the a^2s on the left-hand side by subtracting.
ab	$= c^2 - a^2$	Now divide both sides by a.
$\dfrac{ab}{a}$	$= \dfrac{c^2 - a^2}{a}$	Notice the as on the left-hand side of the equation cancel.
b	$= \dfrac{c^2 - a^2}{a}$	This is the answer.

Notice that the right-hand side can be rewritten:
$$\frac{c^2 - a^2}{a} = \frac{c^2}{a} - \frac{a^2}{a}$$

Here, the as on the right-hand side cancel, so an alternative but equally true answer is:
$$\frac{c^2}{a} - a$$

Formulae involving fractions

When you have to rearrange equations containing fractions, this can involve cancelling the fractions. It is important that you understand what is happening. To make it clear, look at this equation with numbers.

$\frac{6}{3} = 2$ Clearly this is true. 6 divided by 3 is 2.

(?) *What would happen if we multiplied both sides by 3?*

$3 \times \frac{6}{3} = 2 \times 3$ Now the 3s on the left-hand side cancel. The equation is still true.

$6 = 2 \times 3$ Multiplying by the denominator gets rid of the fraction but still keeps the equation balanced.

◎ *Make up some number equations like this for yourself. Be sure that you are clear about what is happening and why it is happening.*

The same principle works in algebra. There are two types of question involving fractions. In the first type, x is in the numerator. In the second type, x is in the denominator.

Example 1

Make x the subject of the formula $\frac{x}{m} = k$

$\frac{x}{m}$ $= k$ Multiply both sides by the denominator, m.

$m \times \frac{x}{m}$ $= k \times m$ This means that the ms on the left-hand side of the equation cancel.

x $= km$

Example 2

Make x the subject of $\frac{j}{x} = h$

$\frac{j}{x}$ $= h$ Multiply both sides by the denominator, as in example 1.

$x \times \frac{j}{x}$ $= h \times x$ The xs cancel on the left-hand side.

j $= hx$ x is now in the numerator on the right-hand side.

$\frac{j}{h}$ $= \frac{hx}{h}$ Now divide both sides by h. The hs on the right-hand side cancel.

$\frac{j}{h}$ $= x$ This is the final answer, because x is on its own.

! **REMEMBER**
Make sure you show your working. It is important to let the examiner see how you have worked out the answers.

Practice questions

Make x the subject.

1) $a + x = c$

2) $x - c = a$

3) $x - m = a + c$

4) $x + h = 2c$

5) $jx = t$

These questions are harder.

6) $\frac{x}{(a + b)} = r$

7) $\frac{x}{d} = (k + 3)$

8) $\frac{j}{x} = \sin 30°$

9) $\frac{m^2}{n^2} = \frac{c^2}{x}$

10) $\frac{t}{x} = \tan 53°$

BITESIZEmaths

Formulae with x^2 and negative x terms

Dealing with x^2

Some students find formulae with the term x^2 difficult. You might find it helpful to remember that x^2 is $x \times x$. That means $\sqrt{x^2}$ must be x.

Example 1

Make x the subject of: $nx^2 = j$

nx^2	$= j$	Divide both sides by n.
$\frac{nx^2}{n}$	$= \frac{j}{n}$	The ns on the left-hand side cancel.
x^2	$= \frac{j}{n}$	Now we take the square root of both sides.
$\sqrt{x^2}$	$= \sqrt{\frac{j}{n}}$	
x	$= \sqrt{\frac{j}{n}}$	We cannot work out the right-hand side any further, so we leave the answer as this equation.

Dealing with negative terms

Example 2

Make x the subject of: $c - mx = y$

$c - mx$	$= y$	Multiply throughout by -1.
$-c + mx$	$= -y$	Now rearrange this.
$mx - c$	$= -y$	Add c to both sides.
$mx - c + c$	$= -y + c$	This now needs tidying up.
mx	$= c - y$	It is neater to rearrange the right-hand side term, so the positive term comes first. ($c - y$ is the same as $-y + c$.) Now divide both sides by m.
$\frac{mx}{m}$	$= \frac{c}{m} - \frac{y}{m}$	The ms on the left-hand side of the equation cancel out.
x	$= \frac{c}{m} - \frac{y}{m}$	Notice, this can also be written as $x = \frac{c-y}{m}$

Practice questions

Make x the subject.

1) $mx^2 = h$

2) $x^2y = (m + n)$

3) $x^2 - b^2 = a^2$

4) $v^2 - y = a(b + x)$

5) $m(x + y) = n$

6) $\frac{(a - b)}{x} = y$

7) $d = \frac{t}{x}$

8) $\sin 45° = \frac{y}{x}$

ℹ️📺Simultaneous linear equations

In this section we will look at solving simultaneous linear equations algebraically. The TV series looks at solving these equations graphically.

Simultaneous means 'at the same time'. When you are solving simultaneous linear equations, there are two equations that you have to solve at the same time. The problem is that with two variables you cannot solve an equation in one go. You have to eliminate one of the letters first and find the value of the other letter, then substitute back to find the value of the letter eliminated.

> **REMEMBER**
> Always set out simultaneous equations in the way shown here. This helps you and the examiner to keep track of your working.

Example 1

Solve: $x + y = 7$ (equation 1)

$x - y = 1$ (equation 2)

In equation 1, if x was 1, y would then be 6. If x was 2, y would be 5, and so on. Eliminating one of the variables is vital.

We need to decide which letters can be eliminated. What happens if we subtract equation 2 from equation 1?

$x - x = 0$ and $y - - y = 2y$

First, we get rid of the xs. Then, we can find the value of the ys.

(?) *What happens if you add the two equations?*

By adding the equations, you eliminate the ys.

$x + x = 2x$ and $y + -y = 0$

> **REMEMBER**
> It doesn't matter which letter you eliminate first, the other letter will be worked out afterwards.

Method 1, eliminate the xs:

$x + y \quad = 7$ (equation 1)

$x - y \quad = 1$ (equation 2)

Subtract equation 2 from equation 1:

$2y \quad\quad = 6$

$\frac{2y}{2} \quad\quad = \frac{6}{2}$

$y \quad\quad = 3$

Method 2, eliminate the ys:

$x + y \quad = 7$ (equation 1)

$x - y \quad = 1$ (equation 2)

Add equations 1 and 2:

$2x \quad\quad = 8$

$\frac{2x}{2} \quad\quad = \frac{8}{2}$

$x \quad\quad = 4$

(?) *What do you need to do now to solve the equation?*

Substitute for y into equation 1:

$x + 3 \quad\quad = 7$

$x + 3 - 3 \quad = 7 - 3$

$x \quad\quad = 4$

Substitute for x into equation 1:

$4 + y \quad\quad = 7$

$4 + y - 4 \quad = 7 - 4$

$y \quad\quad = 3$

(?) *How could you check your answer?*

Check: substitute for x and y into equation 2:

$4 - 3 \quad\quad = 1$ ✓

$x = 4, y = 3$

Substitute for x and y into equation 2:

$4 - 3 \quad\quad = 1$ ✓

$x = 4, y = 3$

BITESIZEmaths

Example 2

You may have to carry out another operation before you can eliminate one of the letters.

Solve: $3x + 2y = 41$ (equation 1) $x + y = 16$ (equation 2)

Here, adding gives us $3x + x = 4x$ and $2y + y = 3y$.

Subtracting gives us $3x - x = 2x$, $2y - y = y$.

So simply adding or subtracting won't work, because we still have two letters.

We need to use another technique – multiplying equation 2 by either 2 or 3. If we multiply by 3 we'll get $3x$ and we can eliminate the xs. If we multiply equation 2 by 2, we get $2y$, so we can eliminate the ys.

Method 1

$3x + 2y = 41$ (equation 1)

$x + y = 16$ (equation 2)

Multiply equation 2 by 2:

$2x + 2y = 32$ (equation 3)

Subtract equation 3 from equation 1:

$3x + 2y = 41$ (equation 1)

$2x + 2y = 32$ (equation 3)

$x = 9$

Substitute for x into equation 1:

$3 \times 9 + 2y = 41$

$27 + 2y = 41$

$27 + 2y - 27 = 41 - 27$

$2y = 14$

$\frac{2y}{2} = \frac{14}{2}$

$y = 7$

Method 2

$3x + 2y = 41$ (equation 1)

$x + y = 16$ (equation 2)

Multiply equation 2 by 3:

$3x + 3y = 48$ (equation 3)

Subtract equation 3 from equation 1:

$3x + 2y = 41$ (equation 1)

$3x + 3y = 48$ (equation 3)

$-y = -7$

Multiply both sides of the equation by -1:

$y = 7$

Substitute for y into equation 1:

$3x + 2 \times 7 = 41$

$3x + 14 = 41$

$3x + 14 - 14 = 41 - 14$

$3x = 27$

$\frac{3x}{3} = \frac{27}{3}$

$x = 9$

! REMEMBER Always check the answer by substituting into the second equation. This will tell you if you have made an error.

To check your answer, substitute the numbers for x and y into equation 2.

Substitute for x and y into equation 2:

$9 + 7 = 16$ ✓

$x = 9$, $y = 7$

Substitute for x and y into equation 2:

$9 + 7 = 16$ ✓

$x = 9$, $y = 7$

Practice questions

Solve these equations.

1) $2x + 3y = 17$
   $x + 3y = 13$

2) $x + 3y = 20$
   $x - 2y = -10$

3) $2x + 5y = 24$
   $5x + 5y = 45$

4) $2m + 5n = 29$
   $m + 3n = 17$

5) $2t + 4v = 26$
   $5t - 2v = 29$

6) $10b + 3c = 36$
   $2b - c = 4$

7) $12w + 7y = 109$
   $5w + 3y = 46$

8) $2m + 3n = 45$
   $m + 5n = 47$

9) $4x + 3y = 69$
   $2x + y = 33$

Using simultaneous linear equations to solve problems

Solve each of these problems by forming a pair of simultaneous linear equations.

1) Find two numbers with a sum of 20 and a difference of four. (Let the numbers be x and y.)

2) Twice one number added to 3 times another number gives an answer of 32. Find the numbers if the difference between them is 1.

3) The line $y = mx + c$ passes through (2,3) and (4,5). Find the values of m and c. (Hint: put the values of 2 and 3 in for x and y

in one equation and the values 4 and 5 for x and y in the other equation. Then solve them simultaneously.)

4) The line $y = ax + b$ passes through the point (3,7) and (5,11). Find a and b.

5) The wage bill for five men and three women is £2250. The wage bill for eight men and ten women is £4900. Find the wage for a man and a woman.

Exam-style questions on equations

1) Janita buys 5 fencing posts and 25 litres of paint for £40. Mark buys 6 fencing posts and 10 litres of paint for £18. How much is 1 litre of paint?

2) Factorise $2bc - b$.

3) Factorise fully the expression $2\pi r + \pi r^2 h$.

4) Make c the subject of this formula:
   $x = b(t - c)$.

5) Solve: $x + y = 20$
   $2x + 5y = 58$

6) Rearrange the formula $v^2 = u^2 + 2as$ to make a the subject of the formula.

7) Make x the subject of the formula
   $m^2 = x^2 + y^2$.

8) Solve: $3x + 2y = 49$
   $x + 5y = 25$

9) Solve: $2a + 4b = 34$
   $5a - b = 41$

This section is about:

- working out probabilities

- finding the probability of two events

- calculating the probability of mutually exclusive events

- the 'OR' rule

- calculating relative frequency

- the 'AND' rule

- tree diagrams

Probabilities are usually worked out in fractions, because they are on a scale from 0 to 1. Make sure you can cancel fractions, because it is important to give your answer in its lowest possible terms.

You also need to be clear about the connection between percentages and fractions. A percentage is simply a fraction times 100. Per cent means per one hundred. Try thinking of the 100 cents that make up a dollar to help you remember this.

You need to be able to change a percentage, such as 90%, to a fraction ($\frac{9}{10}$). 50% as a fraction is $\frac{1}{2}$ (see Factzone).

Make sure you cover the ideas behind relative frequency and do some probability experiments. (Your teacher or college lecturer will be able to advise you on appropriate experiments.)

Everyone has an instinctive understanding of chance. Children, for example, instinctively know when they have gone too far with their parents. Yet as we grow up, we often lose this understanding. Convincing ourselves that a certain horse is a 'dead cert' to win a race is a good example of this. The horse might be well trained and may have won races in the past, but believing it will definitely win in a race tomorrow is silly. It may have a high chance of winning, but there is still a risk that it will lose. The risk of having a crash while flying in an aircraft is extremely low. Air travel is safer than travelling in a car on the motorway, but there is still a small chance that something could go wrong.

You need to understand that probability is a measure of the likelihood that something will happen. It is is not a prediction of what will definitely happen, unless the probability of that event is 1. The probability that the sun will rise tomorrow, for example, is 1.

Probability

A measure of the chance of something happening, on a scale of 0 to 1. Zero is the probability of an impossibility and 1 is the probability of a certainty.

The probability of something NOT happening = 1 minus the probability of it happening.

Random: When something is picked at random it means each thing has an equal chance of being picked. Most calculators now have a random number generator. This is a mechanism for picking numbers for statistical investigations. Find out how this function works on your calculator.

Common fractions and percentages

Make sure you know these common fractions and their percentage equivalents:

$100\% = 1$

$50\% = \frac{1}{2}$

$25\% = \frac{1}{4}$

$75\% = \frac{3}{4}$

$33\frac{1}{3}\% = \frac{1}{3}$

$66\frac{2}{3}\% = \frac{2}{3}$

Adding fractions

If you aren't sure how to add fractions, use the $a\ b/c$ button on your calculator. If you want to enter $\frac{3}{4}$, the button sequence is $3a\ b/c\ 4$ and then the display will show 3 followed by ⌐ and then the 4. This reads as $\frac{3}{4}$. If you want to enter a mixed number, say $4\frac{1}{2}$, enter the 4, press the $a\ b/c$ button and then enter the $\frac{1}{2}$ as before. This will then give you $4 ⌐ 1 ⌐ 2$ in the display.

Multiplication

A convention in mathematics is to use the dot (.) as a multiplication sign. This is often used in probability. It is also very common in the rest of Europe, where the comma represents the decimal point and the dot indicates multiplication. So, in the rest of Europe, 1.3 x 4 is written as 1,3.4.

Probability of a single event

Example 1

If a playing card is picked at random from a pack of 52, what is the probability that it is:

a) a queen   b) the ace of hearts   c) a club?

To answer this, think about the number of cards and the number of chances.

a) There are hearts, clubs, spades and diamonds in a pack of cards so there must be 4 cards that are printed as queen cards. The probability of a queen must be $\frac{4}{52}$. Cancel this down to get $\frac{1}{13}$. The probability of picking a queen is $\frac{1}{13}$.

b) There is only one ace of hearts, so the probability must be $\frac{1}{52}$.

c) There are 13 clubs in a pack of cards, so the probability of picking a club must be $\frac{13}{52}$, which cancels down to $\frac{1}{4}$.

Example 2

A bag contains 6 green balls, 3 yellow balls and 2 red balls. A ball is taken out at random. What is the probability that it is :

a) a red ball?   b) a green ball?   c) a yellow ball?

a) There are 2 red balls out of a total of 11, so the probability of red = $\frac{2}{11}$.

b) There are 6 green balls, so the probability of green is $\frac{6}{11}$.

c) There are 3 yellow balls, so the probability of yellow is $\frac{3}{11}$.

◎   *Add up all the probabilities in the answers. What do you notice?*

They should all add up to 1. You can use this method to check that your answers are correct.

Practice questions

1) A bag contains 8 oranges, 5 apples and 4 bananas. Find the probability that a fruit picked at random is:

a) an orange   b) an apple   c) a banana.

2) The black cards in a pack are removed. One card is selected from the red cards that remain. What is the probability that it is:

a) the king of hearts   b) a diamond   c) not the queen of diamonds?

3) A bag contains 14 white balls, 12 red balls and 12 yellow balls. After 3 white balls, 4 red balls and 7 yellow balls have been removed, what is the probability that the next ball chosen will be white?

4) A fair die is rolled, what is the probability of getting a 6?

5) 'When a fair die is rolled, the probability of getting an even number is not equal to the probability of getting an odd number.' Is this statement true or false? Use a mathematical argument to justify your answer.

⊕Listing outcomes from events

When a die is rolled and a coin is spun through the air simultaneously, we have two events occurring. The result on the die does not affect the result on the coin – the two events are independent of each other. The possible outcomes are:

die	coin	die	coin	
1	h	1	t	h = heads
2	h	2	t	t = tails
3	h	3	t	
4	h	4	t	
5	h	5	t	
6	h	6	t	

An easier way to show these outcomes would be on a sample space. These are also known as possibility spaces or probability spaces. The dots show the combinations of the two events.

score on die

6	•	•
5	•	•
4	•	•
3	•	•
2	•	•
1	•	•
	h	t

face on coin

Probability

Practice questions

1) A 20p coin, a 2p coin and a £1 coin are all spun through the air together. List all the possible orders in which they can land.

2) List all the possible landing orders when a 10 franc coin, a 1 punt coin, a 10p coin and a £2 coin are spun through the air together.

3) A green spinner with the numbers 1, 2, 3 on it and a blue spinner with the numbers 1, 3, 5 on it are spun together. List all the possible outcomes.

4) Marty, Nilesh, Nisha and Kate like to run and race each other. List all the possible orders in which they could finish. In how many races could Nisha finish in front of Kate?

5) A spinner with the numbers 1 to 5 is spun at the same time as a fair die is rolled. List all the possible outcomes that can occur.

BITESIZEmaths

❶Independent events and using the 'AND' rule

If the occurrence of one event is unaffected by the occurrence of another event, then the events are said to be independent, e.g. rolling two fair dice and obtaining a 2 on the first dice and a 6 on the second dice.

The 'AND' rule

❗ REMEMBER The dot means multiply.

When two events are independent, the probability of one event and the other event occurring is the product of the two events. This is the 'AND' rule:

P(X AND Y) = P(X) . P(Y)

Example 1

Two coins are spun through the air at the same time. Find the probability of obtaining a head and a head.

P(Head AND Head) = P(H) . P(H)

$$= \frac{1}{2} \times \frac{1}{2}$$

$$= \frac{1}{4}$$

Example 2

A coin is spun through the air and a fair dice is rolled. What is the probability of:

a) obtaining a head on the coin?

b) obtaining a 6 on the dice?

c) obtaining a head on the coin AND a 6 on the dice?

a) The probability of a head on the coin = $\frac{1}{2}$.

b) The probability of a 6 on the dice = $\frac{1}{6}$.

c) The probability of a head on the coin AND a 6 on the dice = $\frac{1}{2} \times \frac{1}{6} = \frac{1}{12}$.

Practice questions

1) A card is drawn from a fair pack and a coin is spun through the air. What is the probability that:

a) it is a red card?

b) the coin shows heads?

c) the card is red AND the coin shows heads?

2) A playing card is picked at random from a pack and then replaced. The pack is shuffled before a second card is taken. What is the probability that:

a) both cards are spades?

b) both cards are queens?

c) both cards are not picture cards?

⬤⬤⬤⬤ *i* 📺 Using tree diagrams

Tree diagrams are a useful way of demonstrating all the possible outcomes that can come from making choices.

Example 1

A box contains 5 blue discs and 3 red discs. A disc is selected at random and replaced. A second disc is then selected at random. What is the probability that both discs are blue?

To find the probability of a blue followed by a blue, read along the top branch. The probability of a blue and a blue = $\frac{5}{8}$ x $\frac{5}{8}$ = $\frac{25}{64}$. You can find the probability of other possible combinations in this way.

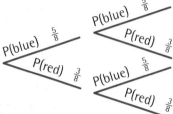

b Example 2

A box contains 5 blue discs and 3 red discs. A disc is selected at random and NOT replaced. A second disc is then selected at random. What is the probability that both discs are blue? What is the probability that one is blue and one is red?

The probability that both discs are blue = $\frac{5}{8}$ x $\frac{4}{7}$ = $\frac{20}{56}$ = $\frac{5}{14}$.

The probability that one is blue and one is red is the same as the probability of a red and a blue, or a blue and a red. This is an example that uses the AND as well as the OR rules.

P(a red and a blue) or P(a blue and a red) = $(\frac{3}{8}$ x $\frac{5}{7})$ + $(\frac{5}{8}$ x $\frac{3}{7})$ = $\frac{15}{28}$.

Notice all the fractions at the end of the branches should add together to give an answer of 1, e.g. $\frac{4}{7}$ + $\frac{3}{7}$ = 1.

Example 3

A bag containing 10 blue balls and 7 white balls is placed on a table. Draw a tree diagram to show all possible combinations for two selections where the balls are replaced after each selection.

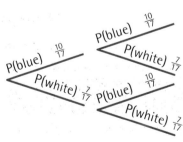

What would the tree diagram look like if the process was repeated with 5 extra red balls added to the bag, and if the ball was not replaced after the first selection?

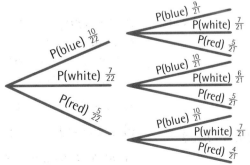

BITESIZEmaths

Probability

1) The number 6790 is multiplied by 10. What is the value of the 9 in the answer?

2) A television is priced at £425. John pays cash and is given a discount of 15%. Calculate the amount that John pays.

3) $\frac{7}{10}$ of the cost of a book goes to the shop where it is sold. A book costs £12. How much does the shop get?

4) Judith is thinking of a number. It is less than 20, it is a multiple of 3 and of 2 and one of its factors is 9. What is the number?

5) Copy and complete this number pattern: 1,1,2,3,5,8,_,_,_

6) Write down all the prime numbers greater than 1 but less than 10.

7) A pattern begins: 1,3,6,10,_,_,_

Copy and complete the pattern. What is this sequence better known as?

8) Solve the equation $3(x - 2) = 15$

9) A bag contains 5 red discs and 3 blue discs. What is the probability of picking a red disc at random?

10) Calculate the area and circumference of a circle with radius 30 m. Give each answer to the nearest whole number.

These questions are for students working at Intermediate level and above.

11) An escalator moves between the first floor and the ground floor in a department store.

a) Calculate the distance XY.

b) Calculate the angle that the escalator makes with the horizontal XY.

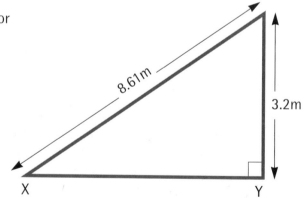

12) A regular hexagon is shown below. One side of the hexagon has been extended to form angle y. Work out the size of angle y.

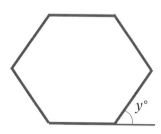

13) The diagram shows a square and a rectangle. The square has sides of $3m$ metres and the rectangle has a length of $2m$ metres and width of 24 metres.

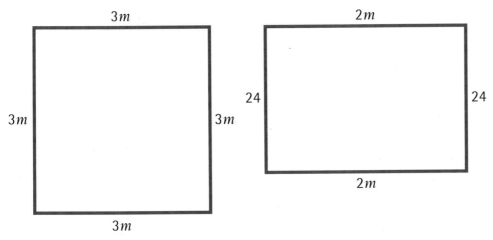

a) Write down an expression for the perimeter of the square.

b) The perimeter of the rectangle is $(4m + 48)$ metres. Explain how this formula is derived.

c) The perimeter of the square and the perimeter of the rectangle are equal. Work out the value of m.

14) In the triangle ABC, angle A = 90°, AC = 24 cm and AB = 7 cm.

a) Calculate the length of BC.

b) Calculate angle C. Give your answer to the nearest degree.

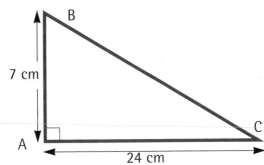

15) Factorise $2bc - b$

16) Factorise fully $\pi r^2 + \pi rh$

17) Multiply out $(3x-4)(x + 5)$ and simplify your answer.

18) Freddie measures the length and width of her classroom floor. It is rectangular, with a length of 8.74 m and a width of 6.26 m. Calculate the area of carpet needed for the classroom, giving your answer to an appropriate degree of accuracy.

19) Find the volume of a cylinder of base radius 6 cm and height 11 cm.

Exam-style questions

Answers to practice questions

Using number

Using number (p13)

1) 320
2) £40.80
3) 16:15 hrs
4) £5.80
5) £1, 20p, 5p, 2p
6) 30 hours, 5 mins
7) 2025
8a) 4 hours 5 mins
8b) 50 mins
9) 9 and 4
10) 9 and 7

Standard form (p13)

1) 2.5×10^2
2) 5.65×10^2
3) 5.6×10^4
4) 5.0×10
5) 3.6×10^3
6) 2.7×10^2
7) 5.8×10^4
8) 2.9×10
9) 8.5×10^5
10) 3.65×10^3
11) 8×10^{-4}
12) 3×10^{-5}
13) 5.5×10^{-4}

14) 3.67×10^{-1}
15) 7.7×10^{-5}
16) 9×10^{-2}
17) 16 zeros
18a) 6×10^{17}
18b) 1.5×10^{-7}
19) 1.333×10^9
20) 1.369×10^{-11}
21) 7×10^3
22) 1.5×10^{-1}
23) 4.57×10^{-19}
24) 5 zeros
25) 6×10^7
26) 1.96×10^{12}

Percentages and fractions

Working with fractions 1 (p17)

1) $\frac{3}{6}$
2) $\frac{6}{9}$
3) $\frac{4}{5}$
4) $\frac{3}{12}$
5) $\frac{3}{4}$
6) $\frac{2}{9}$
7) $\frac{15}{21}$
8) $\frac{14}{16}$
9) $\frac{5}{6}$
10) $\frac{3}{4}$

11) $\frac{10}{3}$
12) $\frac{21}{4}$
13) $\frac{45}{7}$
14) $\frac{19}{9}$
15) $\frac{73}{8}$
16) $\frac{17}{5}$
17) $\frac{90}{11}$
18) $\frac{17}{4}$
19) $\frac{5}{2}$
20) $\frac{113}{10}$
21) $2\frac{1}{3}$
22) $1\frac{1}{2}$

23) $1\frac{1}{4}$
24) $1\frac{4}{5}$
25) $2\frac{2}{3}$
26) $1\frac{4}{8} = 1\frac{1}{2}$
27) $2\frac{1}{5}$
28) $1\frac{3}{6} = 1\frac{1}{2}$
29) $1\frac{3}{12} = 1\frac{1}{4}$
30) $2\frac{5}{8}$

Working with fractions 2 (p19)

1) $\frac{7}{12}$
2) $\frac{3}{20}$
3) $\frac{5}{9}$
4) $1\frac{1}{2}$
5) $2\frac{24}{45} = 2\frac{8}{15}$
6) $11\frac{5}{6}$
7) $2\frac{17}{30}$
8) $\frac{191}{360}$

Measuring

Measuring to the nearest unit (p23)

1a) 1 m

1b) 30 feet

1c) 12 cm

1d) 0.25 cm

2a) 3 feet

2b) most people would probably round this to 120 inches

3) About 26 pounds

4) About 60 cm

5) About 8 cm

Calculating the area of shapes (p25)

1) 17.5 m²

2) 206 m²

3) 3 rolls

4) 7 tins

Finding the volume (p27)

1) 60 m³

2) 108 m³

3) 3770 cm³

4) 0.62 m³

5) 78 m³

6) 49 763 cm³ (49 800 cm³)

Shape and space

Using angle facts (p31)

1) $x = 28°$

2) $x = 30°$, $2x = 60°$, $3x = 90°$

Using Pythagoras' rule (p33)

1) 10.01 cm (to 2 dp)

2) 23.82 cm (to 2 dp)

3) 10.63 cm (to 2 dp)

4) 13.23 cm (to 2 dp)

Using Pythagoras' rule to solve problems (p35)

1) 10.77 cm (to 2 dp)

2) 10.61 cm (to 2 dp)

3) 4.58 m (to 2 dp)

4) 50 km

5) 30 km

6) Prakash

7) 2.97 m (to 2 dp)

8) 22.36 cm

9) 32.02 cm

10) 4.65 m

Data handling

Pie charts (p45)

1)

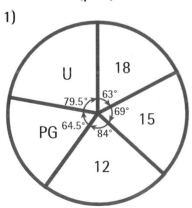

2)

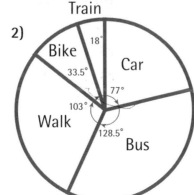
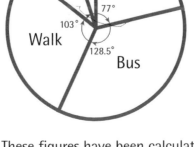

3)

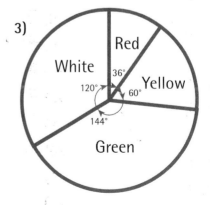

These figures have been calculated to the nearest half degree.

PG

BITESIZEmaths

Data handling continued

Scatter diagrams and correlation (p47)

1a)

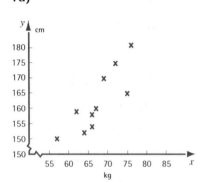

1b) There is a strong positive correlation between height and weight.

1c) Nisha's height will be about 152 cm.

2a)

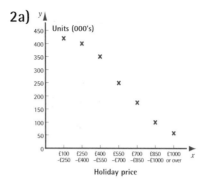

2b) There is a strong negative correlation.

Finding averages (p49)

a) Mean value is 9 apples (9.05). Modal class is 0–4.

b) Mean value is 5.7 cm. Modal class is $1 \leq h < 4$.

c) Mean value is 161.51 cm Modal class is $159.5 \leq h < 164.5$.

d) Mean value is 69.5 marks. Modal class is 55–69.

e) Mean value is 4.975 papers/magazines. Modal class is 0–2.

Cumulative frequency (p51)

1)

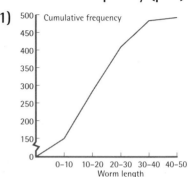

2)

Age	No of people (Millions)	Cumulative freq
under 10	16	16
10–19	12	28
20–29	17	45
30–39	16	61
40–49	15	76
49–69	10	86
69–89	4	90

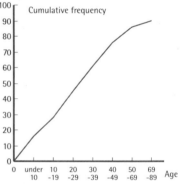

Algebra

Number patterns (p55)

1) Square numbers are a sequence of numbers that are built up by taking each number in the number line and squaring it. E.g. $1^2 = 1$, so 1 is square; $2^2 = 4$, so 4 is square, and so on.

Triangular numbers are a sequence made up as follows: 1, 3 (= 2 + 1), 6 (= 3 + 2 + 1), 10 (= 4 + 3 + 2 + 1), and so on.

Square numbers make a square pattern when drawn out, whereas triangular numbers make a triangular pattern.

2a) Cube numbers

2b) It is built up as follows:
1 x 1 x 1 = 1

2 x 2 x 2 = 8

3 x 3 x 3 = 27 etc.

2c) 125, 216

3a) 15, 21, 28

3b)

3c) Triangular numbers

Sequences (p57)

1) 32

2) 39

3) 25

4) 81

5) 0.038

6) 1250

7a) 19

7b) 8

7c) Multiply by 5 then add 2

8a) 121

The first answer of 4 was generated by 1 x 3 +1

The second answer of 13 was generated by 4 x 3 + 1

The third answer of 40 was generated by 13 x 3 + 1

Algebra continued

8b) 5, 8

9) 47

Solving equations with an unknown number on one side (p65)

1) $x = 4$

2) $x = 10$

3) $x = 20$

4) $x = -15$

5) $x = 3$

6) $y = -2$

7) $y = 9$

8) $w = 60$

9) $x = 7$

10) $x = 4$

11) $x = 4$

12) $x = 2$

13) $x = 9$

14) $y = 1$

15) $x = 5$

16) $x = 8$

17) $y = 2$

18) $x = 3$

19) $y = 2$

20) $x = \frac{1}{4}$

Solving equations with u nknown numbers on both sides (p66)

1) $x = 2\frac{1}{2}$

2) $x = 9$

3) $x = 14$

4) $x = 4$

5) $x = 1$

6) $x = 3$

7) $y = 10$

8) $x = 3$

9) $x = 6$

Solving equations with brackets (p69)

1) $x = 10$

2) $x = 8$

3) $x = 7$

4) $x = -20$

5) $x = 7$

6) $x = 17$

7) $x = 3$

8) $x = 7$

9) $x = 6$

10) $x = 5$

11) $x = 3$

12) $x = 3$

13) $x = 8$

14) $x = 2$

15) $x = 10$

16) $y = 12$

17) $z = 15$

18) $x = 5$

19) $m = 2$

20) $x = 6$

Solving equations with fractions (p71)

1) $x = \frac{1}{2}$

2) $x = \frac{4}{7}$

3) $x = 2\frac{1}{2}$

4) $x = \frac{12}{8} = 1\frac{1}{2}$

5) $y = 2\frac{1}{3}$

6) $t = \frac{9}{12} = \frac{3}{4}$

7) $m = \frac{15}{10} = 1\frac{1}{2}$

8) $a = \frac{5}{20} = \frac{1}{4}$

9) $h = 2$

10) $x = \frac{6}{9} = \frac{2}{3}$

11) $x = 38$

12) $t = 28$

13) $m = 128$

14) $f = 36$

15) $z = 165$

16) $n = 15\frac{1}{6}$

17) $x = 13\frac{1}{3}$

18) $x = 30$

19) $x = 21$

20) $14\frac{2}{5}$

Solving problems using equations (p71)

1) $y = 7$

2) 9 cm

3) The angles are 90°, 18.75° and 71.25°

4) 152, 153, 154

5) $y = 25°$

Factorising (p73)

1) $2(x + 2y)$

2) $3(3b + 5c)$

3) $6(2a + b)$

4) $4(m + 2n)$

5) $5(3b + 4c)$

6) $3(3v - 4w)$

7) $6(v - 2u)$

8) $5(5r + 6s)$

9) $12(2t + 3r)$

10) $27(m - 3n)$

11) $4(3a + 2b)$

12) $3(5x - 8y)$

13) $27(2c - d)$

14) $7(5m - 4t)$

15) $4(3k - 4l)$

16) $8(5h - 3j)$

17) $3(5a - 4b - 2c)$

18) $8(x^2 + 2xy + 3y)$

19) $5(3x^2 - 7y^2)$

20) $32p(2p - 1)$

21) $(x+y)(x-y)$

22) Not a difference of two squares and contains no common factors.

23) $(7t - 8m)(7t + 8m)$

24) Not a difference of two squares and contains no common factors.

25) Not a difference of two squares but it does factorise: $4(3m^2 - 4n^2)$

26) Not a difference of two squares and contains no common factors.

27) $(10k + 9m)(10k - 9m)$

28) $(13c + 15d)(13c - 15d)$

Algebra continued

Changing the subject of a formula (p75)

1) $x = c - a$
2) $x = a + c$
3) $x = a + c + m$
4) $x = 2c - h$
5) $x = \frac{t}{j}$
6) $x = r(a + b)$
7) $x = (k + 3)d$
8) $x = \frac{j}{\sin 30°}$
9) $x = \frac{c^2 n^2}{m^2}$
10) $x = \frac{t}{\tan 53°}$

Formulae with x^2 and negative x terms (p76)

1) $x = \sqrt{\frac{h}{m}}$
2) $x = \sqrt{\frac{(m + n)}{y}}$
3) $x = \sqrt{(a^2 + b^2)}$
4) $x = \frac{v^2 - y}{a} - b$
5) $x = \frac{n}{m} - y$
6) $x = \frac{a - b}{y}$
7) $x = \frac{t}{d}$
8) $x = \frac{y}{\sin 45°}$

Simultaneous linear equations (p79)

1) $x = 4$ $y = 3$
2) $x = 2$ $y = 6$
3) $x = 7$ $y = 2$
4) $m = 2$ $n = 5$
5) $t = 7$ $v = 3$
6) $b = 3$ $c = 2$

7) $w = 5$ $y = 7$
8) $m = 12$ $n = 7$
9) $x = 15$ $y = 3$

Using simultaneous linear equations to solve problems (p79)

1) $x = 12$ $y = 8$
2) $x = 7$ $y = 6$
3) $c = 1$ $m = 1$
4) $a = 2$ $b = 1$
5) man gets £300
   woman gets £250

Exam-style questions on equations (p79)

1) £1.50
2) $b(2c - 1)$
3) $\pi r(2 + rh)$
4) $c = t - \frac{x}{b}$
5) $x = 14$ $y = 6$
6) $\frac{v^2 - u^2}{2s} = a$
7) $x = \sqrt{(m^2 - y^2)}$
8) $x = 15$ $y = 2$
9) $a = 9$ $b = 4$

Probability

The probability of a single event (p82)

1a) $\frac{8}{17}$
1b) $\frac{5}{17}$
1c) $\frac{4}{17}$
2a) $\frac{1}{26}$
2b) $\frac{1}{2}$
2c) $\frac{25}{26}$
3) $\frac{11}{24}$
4) $\frac{1}{6}$
5) False, P(even) = $\frac{1}{2}$
   P(odd) = $\frac{1}{2}$

Listing outcomes from events (p83)

1) 20p,   2p,   £1
   20p,   £1,   2p
   £1,   20p,   2p
   £1,   2p,   20p
   2p,   £1,   20p
   2p,   20p,   £1

2) 10f   1p   10p   £2
   10f   1p   £2   10p
   10f   £2   1p   10p
   10f   £2   10p   1p
   10f   10p   £2   1p
   10f   10p   1p   £2
   1p   10f   10p   £2
   1p   10f   £2   10p
   1p   £2   10f   10p
   1p   £2   10p   10f
   1p   10p   £2   10f
   1p   10p   10f   £2
   10p   10f   1p   £2
   10p   10f   £2   1p
   10p   £2   10f   1p
   10p   £2   1p   10f
   10p   1p   £2   10f
   10p   1p   10f   £2
   £2   10f   1p   10p
   £2   10f   10p   1p
   £2   10p   1p   10f
   £2   10p   10f   1p
   £2   1p   10f   10p
   £2   1p   10p   10f

3) G1B1
   G1B3
   G1B5
   G2B1
   G2B3
   G2B5
   G3B1
   G3B3
   G3B5

Probability continued

4)

Marty Nilesh Nisha Kate
Marty Nisha Nilesh Kate
Marty Kate Nisha Nilesh
Marty Nilesh Kate Nisha
Marty Nisha Kate Nilesh
Marty Kate Nilesh Nisha
Nilesh Marty Nisha Kate
Nilesh Nisha Marty Kate
Nilesh Marty Kate Nisha
Nilesh Nisha Kate Marty
Nilesh Kate Marty Nisha
Nilesh Kate Nisha Marty
Nisha Marty Nilesh Kate
Nisha Marty Kate Nilesh
Nisha Nilesh Marty Kate
Nisha Nilesh Kate Marty
Nisha Kate Marty Nilesh
Nisha Kate Nilesh Marty
Kate Marty Nilesh Nisha
Kate Marty Nisha Nilesh
Kate Nisha Marty Nilesh
Kate Nilesh Nisha Marty
Kate Nilesh Marty Nisha
Kate Nisha Nilesh Marty

12 races

5)

Spinner	Dice
1	1
1	2
1	3
1	4
1	5
1	6
2	1
2	2
2	3
2	4
2	5
2	6
3	1
3	2
3	3
3	4
3	5
3	6
4	1
4	2
4	3
4	4
4	5
4	6
5	1
5	2
5	3
5	4
5	5
5	6

Mutually exclusive events (p84)

1) $\frac{12}{13}$

2) 0.15

3) $\frac{5}{6}$

4) $\frac{1}{2}$

5a) 0.13

5b) 0.7

The 'OR' rule (p85)

1) $\frac{1}{3}$

2) $\frac{5}{6}$

3)

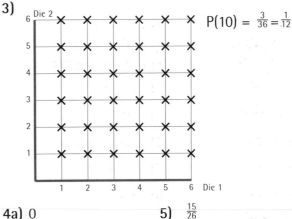

$P(10) = \frac{3}{36} = \frac{1}{12}$

4a) 0

4b) 1, if I want to get to work I will certainly have to use one of these types of transport.

5) $\frac{15}{26}$

Independent events and using the 'AND' rule (p86)

1a) $\frac{1}{2}$

1b) $\frac{1}{2}$

1c) $\frac{1}{4}$

2a) $\frac{1}{16}$

2b) $\frac{1}{169}$

2c) $\frac{100}{169}$

Answers to exam-style questions

1) Nine hundred

2) £361.25

3) £8.40

4) 18

5) 13, 21, 34

6) 2, 3, 5, 7

7) 15, 21, 28 – Triangular numbers

8) $x = 7$

9) $P(r) = \frac{5}{8}$

10) 2827 m², 188 m

11a) 7.99 m

11b) 21.82°

12) 60°

13a) $12m$

13b) Length x 2 + width x 2 i.e. $2m + 2m + 24 + 24$

13c) $m = 6$ metres

14a) 25 cm

14b) 16°

15) $b(2c-1)$

16) $\pi r(r + h)$

17) $3x^2 + 11x - 20$

18) 55 m²

19) 1244.07 cm³